Thor Heyerdahl was born in Norway in 1914. Educated first as a biologist, he subsequently turned to anthropology. A prodigious explorer, he gained world fame in 1947 when he sailed a balsa-wood raft, the *Kon-Tiki*, from Peru to Polynesia, while the subsequent book he wrote about the expedition has been translated into sixty-seven languages. He has since organized many archeological expeditions and ocean crossings in aboriginal vessels. The author of numerous popular and scientific books, his other titles include *Aku-Aku*, *The Ra Expeditions*, *The Mystery of the Maldives*, *The Pyramids of Tucume* and *Easter Island: The Mystery Solved*.

Thor Heyerdahl was for twenty years a member of the New York Academy of Sciences and the Soviet Academy of Sciences, and was chosen by President Gorbachev as his personal adviser on environmental matters. After many years in Italy he now lives in Tenerife, but returns frequently to Peru, the scene of many of his excavations.

Green Was the Earth on the Seventh Day

THOR HEYERDAHL

ABACUS

An *Abacus* Book

First published in the United States and Canada in 1996
by Random House, Inc
First published in Great Britain by Little, Brown and Company 1997
First published by Abacus 1998

A CIP catalogue record for this book
is available from the British Library.

ISBN 0 349 10987 7

Printed and bound in Great Britain by Clays Ltd, St Ives plc

Abacus
A Division of
Little, Brown and Company (UK)
Brettenham House
Lancaster Place
London WC2E 7EN

To my children:

Thor

Bamse

Anette

Marian

and Bettina,

who will help their children

to preserve a green world

CONTENTS

FATUHUKU

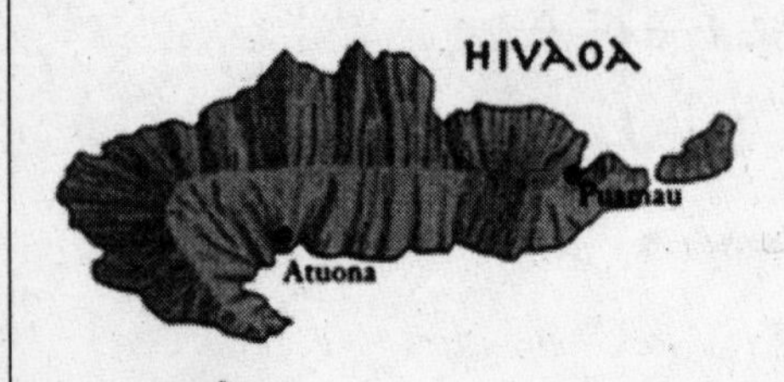

P A C I F I C   O C E A N

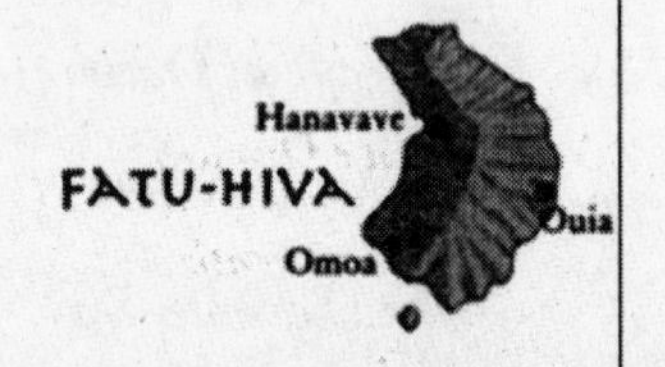

Green Was the Earth on the Seventh Day

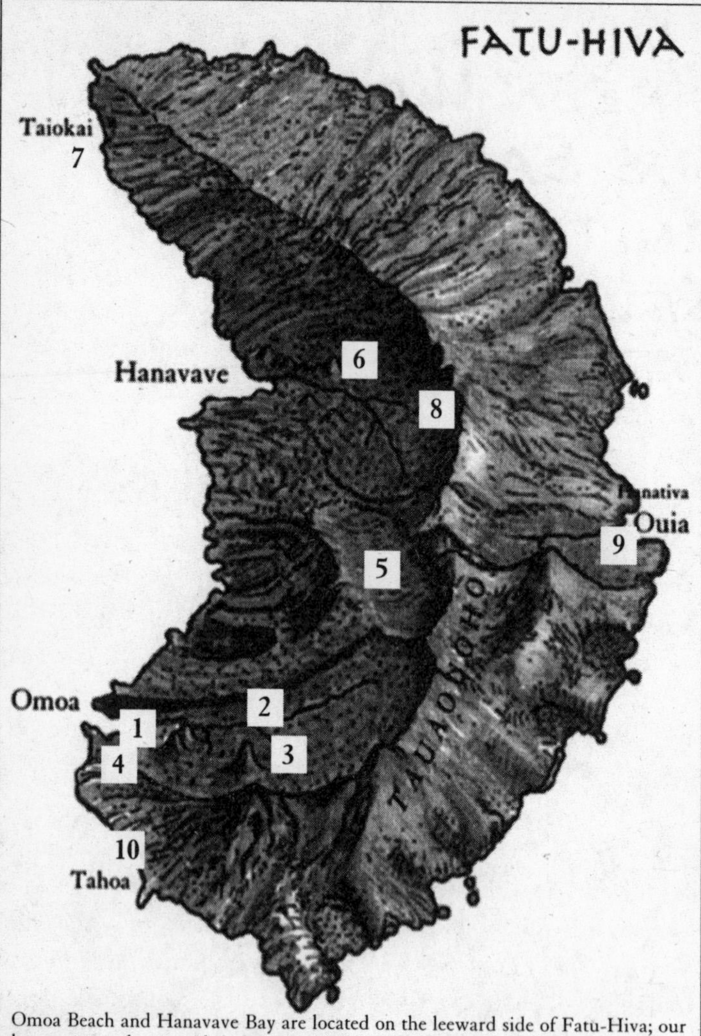

Omoa Beach and Hanavave Bay are located on the leeward side of Fatu-Hiva; our home was on the east side, where the island faces the eternal swells from the Pacific. There is open sea between Fatu-Hiva and Hivoa, which are far larger than the other islands.

1 The beach where we were set ashore. The village in the woods.
2 The bamboo hut on the royal terrace.
3 Rock carvings.
4 Temple ruins and skulls.
5 Unknown highland plateau.
6 Taboo-ridden jungle.
7 "The water of the night."
8 Temple ruins, gods, and statues.
9 Our hut at the cannibal's site.
10 The cave, our last home.

I

Living with a Lost Civilization

PARADISE, FACT OR FICTION? Dream or reality? Lost behind us or tantalizingly ahead?

The Vikings hoped to wake after death, sword still in hand, in a warm Paradise, filled with sunshine and pretty maidens. The Arabs were promised a cool Paradise, with plenty of shade and a river of wine. The Christian Paradise was, until recently, suspended weightlessly among angels above the clouds. Failing to find the angels where they were supposed to be, those who have turned their faith to modern technology have begun to dream of a future Paradise in the company of robots on manmade metal platforms in outer space.

The biblical Paradise was a tropical garden on our own planet, ready-made to feed and nurture newborn mankind in a world filled with plants and animals. It was something man had left and lost in the past. This version, put in writing millennia ago, was saved by the Hebrews, who brought it along on their migrations away from the oldest civilizations in Mesopotamia. It is the earliest recorded version of human history. Ever since man began to build civilization, Jews, followed by Christians and Muslims, have learned that there was once a Paradise on Earth. Myth? Fable? If it ever existed, why did early man start on the long and cumbersome road of progress? And if it never existed, how could man survive in those early ages, when he was the youngest child of nature, naked and empty-handed, on a planet where large and small animals, better fitted for the battle for survival than he, filled every niche in the ecosystem? Was there ever a time or place on Earth where man could come of age to walk about unarmed and unprotected, harvesting his food from an environment no man had cultivated?

If the answer is yes, the old allegory of the Garden of Eden is more than a myth. But then we must ask, again, why did our early ancestors leave such a place, and what urged them to wage war against nature? Why our lasting mixture of contempt and fear for the environment that had once made our very existence possible? Was man driven out of the Garden, or did he walk away sure of building something better, while his God was left sitting alone, resting on his Seventh Day, happy and proud of the marvelous world He had created?

These were the questions planted in my mind during early childhood by a father who trusted God and a mother who believed in Darwin. They were questions that grew to become major issues in my life, that made me leave my parents in a peaceful Norwegian town to try to locate a corner of the long-lost Paradise on Earth. And made me spend much of my later life among peoples still closer to nature than to the civilization we try to sell them.

The question of whether we were heading for – or coming from – a better world has always fascinated me. Made me devote a lifetime of research into cultures of the distant past. As an explorer accustomed to taking bearings where nothing can be seen ahead, I have learned to look at the wakes and the tracks behind to ascertain the proper course. As yet there is nothing to be sure of in the future, but much to learn from the past.

When I put down my pencil and look up from the yellow pages of the manuscript I am writing, I see my own tropical garden with green fields of maize and other food plants inherited from prehistoric Peruvian cultivators, and behind, towering above the banana leaves, their clear yellow outlined against the blue sky, the ancient pyramids of Tucume.

Nobody knows who built them. They were there, abandoned as today, when Columbus crossed the Atlantic and the conquistadores came to Peru. They were there when the Inca armies descended from the high Andes to conquer these extensive coastal plains and to found an empire that included major portions of all the nations that surround modern Peru today.

Peru has always fascinated me. Because of its nature, its history, its amazing prehistory. To the outside world, Peru is best known from posters in tourist offices depicting the huge stone walls of Machu Picchu climbing up the side of a sharp peak soaring into the sky, with jungle-covered ridges behind it. And some may know that Lima is the modern capital on the central coast, while Cuzco, high up in the Andes, was the capital at the time the Spanish conquistadores came to rob the gold and the land from the Incas. But few realize that the modern nation of Peru still is a country five times the size of the United Kingdom, a country which in pre-European time also included all of modern Ecuador, part of Colombia and Brazil, all of Bolivia, and large portions of Argentina and Chile.

Peru faces the Pacific Ocean along its full length, with an open coastline about 1,200 miles long. In the time of the Incas, this coastline was twice that length.

Basically Peru is a tropical country, but blessed by nature with a variety of scenery that can hardly be matched. Snow, desert, and rain forest alternate. The lofty ranges of the Andes split the lowlands in two; desert on the coast, jungle on the inland side. These climatic contrasts at the same latitude are determined by permanent rotations of air and sea in the tropical belt of our planet. The trade-wind clouds blow constantly from the Atlantic side westward across the low jungles of the South American continent until all humidity is squeezed out of them by the lofty Andes, whose glaciers and snowcapped peaks soar as high as 15,000 to 20,000 feet. Between parallel ridges of mountain ranges lie inhabited valleys and highland plateaus 10–12,000 feet above sea level, from which cascades of water flow back east as tributaries to the Amazon river in the eastern lowlands, where the Peruvian rain forests are transformed into the jungles of Brazil.

Few streams find their way in the opposite direction, toward the Pacific coast. So few, indeed, that the Marañón river, with its sources in mountains visible from the Pacific coast, winds eastward and down into the green jungle to cross the entire South American continent and empty into the Atlantic. As a result of the lack of rain, the coastal belt west of the Andes is a vast desert.

It is in this desert I have built my new home. For in the endless stretches of sand and crumbling rock there are some oases. At intervals down the long desert coast there are so-called *quebradas*, dry river beds where water runs in season when there is heavy rain in the highlands. Long before the conquistadores came, more than a thousand years before the Incas descended to the coast, the Mochicas and other long-extinct civilizations improved upon the natural *quebradas*, partly replacing them with canals and elevated aqueducts that run for hundreds of miles across desert valleys and along mountain sides to fork into highly complex systems of large and small irrigation trenches. Where water runs, everything can grow. Many large desert valleys all along the coast were, in pre-Inca times, green fields and gardens that produced a great variety of tropical fruits. Today modern engineers are reopening many of the pre-Inca irrigation canals that have been left dry and abandoned for untold centuries. My own workmen have dug a modest trench a few hundred yards long to link my little oasis to the giant Mochica canal recently reopened by the modern government of Peru. Water from the Andes is again gushing past the Tucume pyramids as it did in prehistoric times, along a canal almost a hundred miles long. Well-preserved prehistoric sections of it are twenty to twenty-five feet wide, wide enough for two good-sized reed boats to have passed each other with whatever cargo they might have brought to or from the sea.

What next to nobody knows about this coastal desert, except the Peruvians themselves and the people who live here, is that at intervals this landscape is turned into South America's largest lake. A lake of brackish water, where pelicans and ducks swim together, and people sail and paddle with balsa rafts, catching fish with their nets. In normal years, Lake Titicaca is the largest lake in the continent. This inland sea lies high up in the Andes, 12,000 feet above sea level. About half of it is inside the southern borders of Peru, the rest extends into neighboring Bolivia, but there are certain abnormal years when the rainwater floods down in the wrong direction from the Andes and overflows the deserts along the north Peruvian coast. Such years are known and dreaded by all Peruvians as 'Niño years,' named after the Niño current off the northern coast. This current changes periodically, causing dramatic climate changes throughout the whole nation. As important in determining the climate of Peru as the permanent easterly trade winds are the two ocean currents that flow like mighty rivers close along the coast. The largest is the Humboldt Current, which carried our *Kon-Tiki* raft to Polynesia. Its cold water sweeps from the Antarctic all the way up along the shores of Chile and Peru almost to Ecuador before it turns west and heads straight to Polynesia. It cools the air and covers Lima and the southern coasts of Peru with a compact cloudy mist half the year. A smaller current, the Niño or 'baby' current, comes in the opposite direction, bringing warm tropic water southward from Panama across the equatorial belt. The two currents collide off the extreme north coast of Peru, and flow side by side westward on either side of the Galápagos Islands, the Humboldt Current green from plankton, the Niño current transparent blue. But in certain times, at intervals often of seven or eight years, the Niño current receives impulses not yet fully understood from a global pattern of disturbances that makes it grow to spectacular dimensions. This usually happens around Christmas – hence the name Niño, referring to the Christ child. In such years the Niño pushes all the way down the coast of Peru, forcing the Humboldt Current to change course and bear directly down upon Easter Island. The warm water kills billions of fish and seabirds and brings famine and catastrophe upon the fishermen who inhabit the north coast. Worse still, it changes the climate; cloudbursts bring heavy rains and violent floods from the mountains, causing disastrous inundations. The worst of recent years was in 1983, when all the coastal lowlands of northern Peru were submerged in water both from the skies and from the ocean. At that time, the American anthropologist James Vreeland counted about 600 balsa rafts bringing fishermen from the north coast, who were catching large quantities of *lisa* fish as they swam in over desert sand and cultivated fields. Tucume could not be reached except by boat, and all the streets were navigable. Coffins sailed away from the cemetery on the current. And the Tucume pyramids, on the high land where I have built my home, emerged as an island in the sea.

Layers upon layers of dried mud, which we find when we dig at the foot of the pyramids, tell us that many such disasters, many Niño years, have caused catastrophes in this part of Peru since earliest prehistoric times.

One such major catastrophe has been dated by Peruvian archaeologists at around A.D. 1100, when floods from a disastrous high tide combined with torrential rains to play havoc the length of the Peruvian coast, destroying local settlements. As a direct result, a year of equally disastrous drought followed throughout the Andean highlands, putting an end to the important Tiahuanaco civilization. This natural catastrophe, which then hit all of Peru, must have influenced all adjacent parts of the Pacific. It is hardly any coincidence that Polynesian historians show that around A.D. 1100 a period of unrest began across Polynesia, with a vortex of inter-island migrations that interrupted the genealogies and changed the royal lines on all the major islands. On Easter Island, archaeologists have found that A.D. 1100 marks the beginning of a new cultural period, during which the local population began to carve their typical long-eared images of stone and wood. The Easter Islanders themselves describe their ancestral fatherland as a desert land far to the east, and their legend says: 'It is a bad land. When the sea is high we die many. When the sea is low we die few.'

Knowing how much damage the Niño flood of 1983 did to Tucume village and the old pyramids, we can imagine that the record Niño year of about 1100 must have been a disaster. The legend narrated here by the first Spanish conquistadores states that it was a terrific flood that put an end to the original Naymlap dynasty that built the Tucume pyramids.

The members of that early civilization were no longer on this coast when the Incas descended from the highlands. At that time, Chimu kingdoms ruled there. They had already conquered and annihilated the civilization that had erected these spectacular monumental pyramids, and not even the Incan historians could give the arriving Spaniards the least information about the origin of these colossal buildings and the people who had constructed them. The Incas had looted the Chimu kingdoms and carried all the gold and treasure up into their own cities in the highlands.

The Spaniards found only poor fishermen and farmers on the coast, and pushed quickly on to get their hands on this fabulous wealth. The pre-Incan road from the coast to the Inca fortress city of Cajamarca went straight by the ruins of Tucume at that time, and two of the early chroniclers referred to them as the most impressive remains of antiquity in the entire Inca empire. But the descendants of the conquistadores built new roads, and the very existence of the mighty pyramids of Tucume passed into oblivion. When I first visited the plains on the northern coast of Peru they were not even marked on any Peruvian map.

I had never heard of Tucume until the day I saw it. Nor had my traveling companion, Guillermo Ganoza, even though he had lived all his life on this north Peruvian coast; he had a real passion for archaeology, and every room in his Trujillo manor house was full of masterpieces of pre-Inca art.

Guillermo had helped focus world attention on the now famous ruins of Chan Chan by inviting the whole board of *National Geographic Magazine* to Peru to see them. When we walked together in between the Tucume pyramids with Walter Alva as our guide, Guillermo exclaimed: 'This is more impressive even than Chan Chan!' And seconded by Walter he challenged me to organize excavations.

Excavations!

We were sitting on a manmade hillock surrounded by pyramids that rose above our heads, one behind the other, as tall as twelve-story buildings.

'There are twenty-six large pyramids and many small ones,' explained Walter. He had walked around all of them and could tell us that what we were seeing were the unexplored ruins of a mighty temple city, whose walled enclosures, ceremonial ramps, and passages, together with the pyramids, covered 220 hectares. And, most important, these abandoned ruins, named 'The Purgatory' by the first arriving Spaniards, were entangled in so much superstition that they had not yet been looted by the tomb robbers who had ransacked all other visible archaeological sites in Peru.

The tomb robbers, or *huaqueros*, who had left the Tucume pyramids in peace had nevertheless played the most important part in the events that had brought me to Tucume. They had played a still more important part in the lives of the two men who had brought me there. The tomb robbers represented the great paradox of Peruvian society. Nobody had, to the same extent as the *huaqueros*, filled the museums and private collections of the world with the splendid masterpieces of pre-Inca art in gold, silver, and ceramic that have made their nation rank with such countries as Egypt, Iraq, and Mexico when it comes to displaying the remarkably high level of pre-European culture. The *huaqueros* are criminals according to the law; they have done irreparable damage to the science of archaeology, they are prosecuted if they are caught, even beaten and shot to death if they try to escape. But Peru and science owe them a great debt. And Peru and science must somehow manage to catch them in a net, haul them out of their obscurity, tame them, train them, and benefit from their outstanding skill, their inherited intuition and experience.

Most of our knowledge about the great coastal civilizations that had been wiped out by the Incas before the advent of the Spaniards we owe to the *huaqueros*.

Professional archaeology is a young science. It was started in the Danish middens in the last century, spread to Norway and the rest of Europe first, and reached Peru only in the most recent decades. The National Museum in Lima and the homes of wealthy Peruvian families display endless collections of pre-Incan art, which almost without exception have come from the *huaqueros*. Until recent years, wealthy Peruvians, knowing it was not entirely permitted by the law, paid professional *huaqueros* to search for ancient treasures, competing among themselves in their exhibitions of the finest pieces and convinced that their country was better served by having its antique art visible and enjoyed above-ground rather than hidden.

I had met Guillermo because I had come to see his stately manor house in Trujillo which, as noted, was full of art bought over centuries from *huaqueros*. And then I traveled northward along the coast with him, because in the desolate ruins of the pre-Inca fortress city of Jequetepeque we met an American archaeologist friend, Christopher Donnan, who told us that a *huaquero* had just been shot dead after looting the greatest gold hoard of the century from a pyramid known as Sipán farther north on the coast. A Peruvian archaeologist had taken over the site, and marvelous treasures had been confiscated before they could disappear from the country.

The Peruvian archaeologist was Walter Alva. It was he who brought me to Tucume. Thanks to the *huaqueros*, I met Walter. I was probably one of the first to visit him in Sipán, before he started excavations and found the 'Lord of Sipán' that would quickly make the site world famous. When I arrived with Guillermo, Walter was at the bottom of a ten-meter-deep looter's pit, tempting me to descend a ramshackle bamboo ladder. Once I got to the bottom, he showed me the various side tunnels the *huaqueros* had dug to locate more tombs. They had started to dig a new pit on the surface, but Walter had been warned; police surrounded the nearby village and confiscated all the treasures that had not yet been sold. One *huaquero* paid with his life when he tried to escape, and Walter and his little team took over the work the looters had interrupted.

Never have I seen a dig like the one initiated by the *huaqueros*. They had dug a square pit vertically down through a hoard of hundreds of ceramic jars in human form, leaving a multitude of little men the size of beermugs in total confusion around the walls, some face forward and some lying backward in the dirt fill. All had the same big disks in their earlobes; all were Long-Ears. To the looters they represented nothing but fill and refuse through which they had to dig on their way down to the sarcophagus they knew they would find deep below.

Walter climbed up from the looters' pit to show us the sort of treasures he had confiscated and brought to the Brüning Museum in the nearby city of Lambayeque, of which he was the director. He stroked his bushy black beard, his eyes justly beaming with pride when he let us see what he had saved for science. It exceeded anything as yet discovered in South America, and I understood immediately that what we were now looking at would change the prehistory of Peru, and perhaps modify our entire concept of pre-Columbus America. These were not vestiges of a primitive, formative culture, yet they were estimated as dating from the very beginning of the Mochica culture, well over a thousand years before Columbus set sail. Centuries before the Vikings ventured into the ocean. When only the great Old World civilizations of antiquity were known to be navigators. What the unfortunate looters had found, but not yet sold, included mummy masks and elegant gold and silver jewelry, inlaid with turquoise, lapis lazuli, and red seashell. Masterpieces of art, with some pieces estimated to be worth $100,000 each on the American black market. But it was not the gold and silver or money value that fascinated me. Here was evidence that these people were expert mariners.

Where did the builders of the Sipán pyramids get their raw materials and their knowledge of working gold and silver? None of the materials needed by the craftsmen to produce these mummy masks, the inlays in their jewelry, or their gorgeous earplugs were available on the Peruvian coast. Gold and silver they could locate, if they knew what to look for, by sending expeditions into the highlands of the Andes.

But the lapis lazuli needed for the incredibly blue eyes of the masks, and all the green turquoise used in the jewelry, could not be found in Peru. The lapis lazuli must have been obtained from Chile – some two thousand miles to the south. Further away than a voyage across the Atlantic. Nobody could walk the entire desert coast of South America, across all the river valleys occupied by other peoples, to locate the mines of lapis lazuli far down the coast of Chile. Such a voyage would only be possible by sea. But the strong Humboldt Current flows northward along the South American coast and makes it next to impossible to sail southward close to land. So the balsa-wood rafts or reed ships, both used by the early Mochicas and represented as a dominant motif in their art, must have sailed south *outside* the wide coastal current and returned with great speed close to land. The raw materials left in their tombs proved that they had been deep-sea voyagers.

The nearest source for all the turquoise would have been in Argentina. During the very centuries when the Mochica kings built their mud-brick pyramids on the coastal plains of north Peru, the priest-kings of Tiahuanaco – and the gigantic Akapana pyramid – dominated the southern shores of Lake Titicaca. And they imported turquoise from quarries in Argentina. Their merchantmen in the highlands had access to the Pacific coast along pre-Incan roads. Tiahuanaco was the cult center of the seafaring sun god and priest-king Kon-Tiki, the pan-Peruvian monarch who, according to Inca traditions, brought culture to South America. It was he who had come as an immigrant, and later left from the coast again with his entourage of fair-skinned and bearded followers, for whom the Europeans were later mistaken. Did the strange traditions of these fair, pre-European visitors give us a possible clue as to why the pre-Inca mummy mask from Sipán had blue eyes? The artists who designed the mask had gone to much trouble to give them these. There was plenty of brown and black stone in Peru, but it took a long journey to fetch blue lapis lazuli from Chile.

I had just come from visits to Arica in northern Chile and Ilo in southern Peru. Modern archaeologists have recently discovered that these two ancient ports are full of tombs containing Tiahuanaco art and trade artifacts, and that both were directly linked to Lake Titicaca and Tiahuanaco by skillfully prepared pre-Inca roads. Obviously, if the early mariners from Sipán and the coastal towns in the Lambayeque valley sailed south to Tiahuanaco ports in Chile, they could bring back both Argentine turquoise and Chilean lapis lazuli.

But not the *spondilus* shell. The teeth of the golden masks from Sipán were of *spondilus*, a tropical seashell that never existed in the waters of Peru, where the Humboldt Current brings cold water from the Antarctic. The *spondilus* found in abundance in the tombs of Sipán could only have come from Panama or Ecuador. When Walter later excavated 'The Lord of Sipán,' he found a large number of these tropical shells placed around the copper sandals of the regent's feet. The link to the sea was obvious.

One conclusion was inevitable. The royalty of Sipán had ruled over a people with an astonishing geographical knowledge. Walter's first radiocarbon date was taken from the roofing beams of the Sipán tomb and came out at about A.D. 290. At that time these pyramid builders of Sipán already knew the entire west coast of South America, were able to secure the blue eyes they wanted for their mummy mask and the *spondilus* shell for its teeth.

Fresh thoughts and new visions circled in my head when Walter saw my enthusiasm and immediately wanted to show me Tucume. The two hours' drive from the twin pyramids of Sipán to the twenty-six awaiting us in Tucume was like bumping softly along on top of the clouds. Everything seemed unreal, like a dream of a visit to another planet, as we walked through the algaroba (carob) forest in the outskirts of Tucume village and the vision of the mountain-size structures opened up before us.

The setting sun painted gold down the slopes of all the pyramids. Ravines caused by centuries, perhaps millennia of erosion ran downhill in parallel grooves as if marked by the claws of a dragon that could be hiding around any corner. Such a sight seemed unreal on our own familiar planet. If such pyramids really existed here and now, it should have been known, if not to me, at least to Guillermo, or to the government people I immediately informed down in Lima. But none of them had seen or even heard of Tucume – only Walter, a few of his colleagues, and the people in the local village (who hardly ever ventured in among the ruins, which to them still were 'El Purgatorio') knew of the place.

I did not know then that in the green belt of algaroba forest and tropical fruit trees that surround the lifeless pyramid city are hidden the modest homes of some twenty *brujos*, or witch doctors. They sometimes assemble in the moonlight among the ruins to communicate with all the supernaturals that other people fear.

Nor did I know then that I would build a home next to the Tucume pyramids, so near to them that the modern world was totally lost from sight except for the greenery around me and the distant backdrop of the Andes.

I was far from my own kin and my own generation. And yet I felt at home here. As if this was the center of the world, of my world. And of the world of the navigating 'Long-Ears,' whose legendary exploits in pre-Inca times had once lured me to drift to sea with my friends on a balsa raft. And who had brought me time and again to Easter Island to tackle the mystery of the long-eared stone giants erected there.

This was the place, according to Inca traditions, where the divine priest-king Kon-Tiki had descended to the lowland on his march through Cuzco and Cajamarca when he forever left his sacred abode in Tiahuanaco. He and his entourage of white and bearded men had passed the pyramids of Tucume on their way to the coast along the same pre-Inca road that, centuries later, brought the Spanish conquistadores from the coast up into the mountains.

Cuzco, which meant 'The Navel of the World' to the Inca, was conquered by a handful of white and bearded Spaniards, because they were mistaken for Kon-Tiki and his followers returning. Easter Island was known as *Te-Pito-o-te-Henua*, or 'The Navel of the World' to the Long-Ears who settled there. And that's what Tucume became to me. And to me the navel cord itself was the Humboldt Current, which linked the coast of Tucume and Sipán to the islands behind the horizon. In these forgotten pyramids lay buried the last secrets of the original Long-Ears. The Lord of Sipán was one of them, buried under hundreds of little ceramic men all of whom had disks in their ear-lobes. And the great lord himself wore splendid gold-and-turquoise ear ornaments so intricate and beautiful in their execution that they have been judged to be perhaps the finest example of pre-Columbian jewelry ever found.

Old traditions and new archaeological finds began to fit rather well together. So well indeed that when the local people told us their legends about the past, the skeletons we dug up from the base and the top of the Tucume pyramids seemed to have their flesh and skin restored when we looked at the local workmen who helped us dig. Nobody doubted that the people still living in this valley were almost pure descendants of the pre-Spanish population. Their healthy sun-tanned faces were as strong and well-shaped as the prehistoric portraits in Mochica pottery. Fragments of the same traditions recorded by the Spanish chroniclers four centuries ago were still alive in their memories.

Only archaeological excavations would tell us the true history of Tucume. Nevertheless it would be unfair and unwise to ignore the oral records that were sacred history to the Peruvian nation when the first Europeans arrived. What the population on the north coast knew to tell about their own ancestry when the Spanish conquistadores came ashore was put into writing independently by two of the early chroniclers. Both were told of a spectacular arrival of ancestral immigrants landing on the beach of Lambayeque, directly below Tucume.

One of them, Father Cabello de Balboa, wrote of his landing on the same beach:

> The people of Lambayeque say – and all the folk living in the vicinity agree with them – that in times so very ancient that they do not know how to express them, there came from the northerly part of Peru, with a great fleet of *balsas*, a father of families, a man of much valour and quality named Naymlap; and with him he brought many concubines, but the chief wife is said to have been Ceterni. He brought in his company many people who followed him as their captain and leader.

We also learn that his court alone consisted of forty men, several of them recalled by name. When he walked ashore in the bay just below the present site of Tucume, his conch-blower walked in front and signaled his arrival by sounding his conch, while a special servant went ahead to strew seashells wherever he was to set his foot. He brought with him on his raft a stone statue, which he set up on the site where he built the huge temple pyramid Chot, still a truly colossal hill towering some three kilometers from the shore. His son Xium became the father of Calla, who founded the temple city of Tucume.

The Naymlap dynasty ruled in the Lambayeque valley for ten generations, until the reign of King Tempellec. He decided to move the ancient stone statue to some other site. Then began the heaviest cloudburst ever seen in these parts. Pouring rain and disastrous floods lasted for thirty days and was followed by a year of severe drought and famine. The priests and the chiefs blamed the king for the disaster, saying it was because he had moved their stone image. They tied his hands and feet and sent him out to sea on a raft. With this king, Tempellec, who drifted away with the Humboldt Current, ends the oral history of Tucume.

Some fragmentary tradition has it that Tucume was named

after a chief named Tucmi in the days of King Callac. Other traditions, still strong, say that the original name of Tucume had been different, and had meant *Lugar de Araña*, 'Place of the Spider'.

Why the spider? This was a mystery that was soon to become meaningful. Walter Alva dug up still another royal tomb in Sipán. The great lord he now found was buried with a most remarkable ornament around his neck: a heavy chain of gold composed of ten large spiders each set on a finely spun spider web of gold filament. The body and eight flexed legs realistically depicted the spider, but each of them had a beautiful human head and the royal symbol of a plumed serpent on their ventral side.

Spiders of pure gold around the neck of a lord in Sipán, and a few miles away in the same Lambayeque valley lay Tucume, once known as the Place of the Spider. Was there some connection between the long-eared royalty of this valley and the Long-Ears who had discovered and settled Easter Island? If so, this would explain why the Easter Islanders, on the coming of the first missionaries, recited texts allegedly learned by heart from their undeciphered written tablets, and referred to their ancestral home in the direction of the sunrise as the Home of the Spiders.

It is indeed 'The Home of the Spiders' I can see from my windows, according to tradition. The Spider Kings were Long-Ears, according to archaeology. I have built my home in a place where the past is still part of the present. I ask myself: What secrets are to be found in these unlooted pyramids?

Other people ask me: Why do you live here? Why not let other people do the digging while you enjoy all the blessings of modern city life?

It is a long story for those who care to know. Parts of it I have told before, with years between as intervals, in tales about my expeditions. And in volumes about the purely scientific results. All this I will reduce to a minimum in the following story. It is the tale of a boy from a small town in Norway who grew up with a passionate love for nature. And with a feeling that modern civilization was on a wrong course. We were building a new Tower of Babel without a blueprint. We were slowly cutting off the branch of the tree of nature of which we were still a part. If God had created nature, we ought to respect it. And if there was no God, then nature was the creator of man, and there was even more reason to respect it.

Nature became to me in early childhood what a church was to many of the adults in my town, and what a workshop or a garage was to most of the boys my own age. I became fascinated with people and cultures living in intimate harmony with nature.

II

Planning to Turn the Clock Backward

NOBODY LIKES TO BE A SLAVE of the watch, but most people like to have one. Time is money. Ancient cultures had plenty of time. A shortage of time is a sign of modern civilization. I can well remember the age when all my friends looked forward to their first watch. I, too, was pleased when I got one. But then came the day when my greatest dream was to turn back time. To get rid of the clock. To try to cut off all ties with civilization and walk into a tropical wilderness empty-handed and barefoot, as a man at one with nature. I looked forward to celebrating that planned event symbolically by putting my watch on a stone and preparing to smash it with another stone, sending cogwheels and springs flying. But the day I got ready to do it, an islander in a loincloth came out of the bush and begged me to give him the watch. So I was deprived of a pleasure I had been looking forward to for years.

From the days when I had barely started to shave, I had the wild idea of going completely back to nature. Completely. To attempt a farewell to life in a civilized community, in any kind of community, with all its good and evil, to study our own civilization from the 'outside'.

I was still a high school boy in Norway when I began to rehearse for this wild adventure. My home was a white-painted wooden house, covered with ivy, in the little coastal town of Larvik at the mouth of the Oslo Fjord. Everything clean. No smog, no pollution. No stress, nothing, apparently, to escape *from*. No hippies or homeless. The largest buildings in the town were a white wooden church on some low cliffs by the harbor, and my father's brick-walled brewery in the central city square. The air was pure and the river clean. It was safe to drink from any running stream. There were rivulets in the forest, the tinkling of tiny waterfalls mingling with the twittering and singing of an infinity of birds.

The water in the harbor was crystal clear, and little boys with long bamboo rods would sit in rows on the wooden pier watching schools of fish wriggling up to nibble at their bait. There was nothing to see at the bottom but the art of nature – smooth boulders, shells and starfish, among gently waving seaweed. At anchor in the fjord lay large whaling ships, with thousands of tons of blubber from the marine giants that still roamed the southern seas in countless numbers. The little city of Larvik was thriving on its wealth of timber and its successful whaling fleet.

But there was something in the air. Modern whaling techniques had made the giant whales such an easy catch that there were real fortunes in the whaling industry. This called for caution. The ocean ran on with no beginning and no end, reaching beyond both poles. But man, too, was everywhere. Could he, with his technical progress, one day bring an end to nature's eternal supply of whales? Today we know, unfortunately that the answer is yes. But that was unthinkable then. The world of the whales was endless. The blue ocean was as infinite as the blue sky, one merging into the other, both part of the boundless universe.

At that time, too, the unrestricted air space had just begun to interest adults. Fairy tales came alive in the minds of boys. Grown men had already begun to lift themselves above the earth like witches on broomsticks or wizards on flying carpets. Along with the other children, I climbed the wobbly tiles of our house to wave and yell as we heard the drone of a single-engined airplane that passed as a speck on the horizon. We even ran to the garden fence to watch the first driver who dared to force his automobile up the steep street past my home. How exciting!

But what a choking smell it left behind, compared with that of a horse.

Traffic in Larvik was still dominated by the rhythmic clacking of horses' hoofs and the rumbling of wheels over cobblestones. In the winter, however, the streets were white-padded and silent, although the tooting of horns had now begun to mix with the familiar jingle of sleigh bells. My father enjoyed the walk back from his office for a peaceful family lunch and a long afternoon nap. The days were never rushed, and watches were big, with ample space between the hours.

Before dusk my father and I often walked down to the piers to inspect the fishermen's catch. Their boxes were filled with lobsters, crabs, shrimps, and an endless variety of fish and other marine creatures, flipping, wriggling, and crawling around and over each other in a fresh odor of seaweed and salt sea. Seafood was a real delicacy for those of us who could afford to get it to the kitchen alive. If delivered dead on ice it was no longer for the discerning.

From my two bedroom windows, upstairs, I had a splendid view of the fjord below and beyond the center of the town, which lay in terraces below us: white gables and red roofs, partly hidden among garden trees where cocks crowed from the backyards. The old seaport of Larvik was like a large but tidy village, climbing from the fjord in a pattern of terraces and ramps up the sides of low forested hills – a verdant landscape of huge pines, firs, oaks, birches, and even Norway's only large beech forest. I could see it all by sneaking out of my bed. In winter, my parents left my bedroom window ajar at night, and I would take only a quick look at the many city lights and the falling flakes of snow before I huddled under the warm eiderdown. But on bright summer nights they would leave both windows wide open, and when they thought I was sleeping, I could sit and dream on the windowsill, high above our terraced garden. I would rush to the window at the distant sound of a ship's bell announcing its departure. As it slid toward the open mouth of the fjord, perhaps to tropical dreamlands hidden from the sight far beyond the hilly headlands that marked the gateway to the open ocean, my soul went with it. Beyond that gate was the boundless, endless world that man was still exploring.

Part of it was unknown. Our compatriot Amundsen had reached the South Pole just before I was born, and now he and others were competing to reach the North Pole by air, since Nansen's drift with the *Fram* across the polar sea had shown that the top of our planet was all floating ice. In the warmer areas, other expeditions were struggling on foot, as in the days of Cortez and Pizarro, to fill in empty spaces on the world map in Africa and South America.

What an enormous world man lived in then! The children in the nearest house below us were emigrating to America with their parents. Weeks of travel beyond the horizon. They were seen off by the rest of us as we would later watch departing astronauts. Certainly nobody would ever hear of them again.

An explorer. That was what I wanted to become. To penetrate on foot, by horse, or by camel unknown parts of our vast world. No ocean had yet been crossed by airplane. Oceans were crossed by steamships. America was only a quarter of the distance away in time compared with Columbus' day, but even that was far. It took three weeks to travel to America. It even took me three days to travel with my parents from the port of Larvik to our mountain holiday cottage in central Norway, partly by train and partly by pony and trap. I was amazed beyond description when some friends showed me what they called a 'radio', a square battery box with holes into which we plugged our earphones and eagerly discussed whether or not we heard music.

A new era was vaguely in the making. My parents received it with pride and excitement. My father trusted the human brain, because it was the gift of God. My mother put all her faith in Darwin's theory of evolution and was confident that man was gaining in intelligence with each generation. Both were sure that man was changing our planet into something ever better. World War I, which had been raging when I was born, would never be repeated. With science and technology, man was marching on toward moral rectitude and peace.

But that was as far as my parents agreed. Their ideas about a possible afterlife collided completely. My father was not too saintly for a little flirtation, but he was convinced God would forgive his sidesteps and let him enter a happy Paradise. My mother was rigidly moral in every sense of the word, but an atheist who believed no more in any god than she did in mountain trolls and mermaids. My father would come tiptoeing upstairs when my mother had put me to bed. He would sit down on the bed while I sat up straight-backed with folded hands to say the Lord's Prayer. It gave both of us a warm and happy feeling. But often we heard the creaking of fast steps coming up the wooden stairs of the eighteenth-century house, and my mother would be standing in the doorway waiting to turn off the light as my father left. Once I heard her ask: 'What kind of nonsense are you teaching the boy!'

I was left alone in the dark room wondering who was right.

Was it meaningful to pray? Sometimes the warmth of my father's simple faith seemed to penetrate my body, and my mother's cold scientific reasoning my brain. Sometimes I was afraid of sleeping alone in the big old house. I dreamed of witches and ghosts, even though I knew they did not exist.

Although I shared my father's love for nature and my mother's passion for zoology and 'primitive' tribes, I could not understand their enthusiasm for modern man's determination to sever all his ties with nature. What was it they wanted to run away from? Were they scared by the ape-man Darwin had painted behind them? They were ready to welcome any change from the world of their own parents by calling it 'progress', no matter what the change might be. 'Progress' was synonymous with distance from nature. The adults, who set the pace, were so absorbed by their own ability to invent and to alter the existing world, that they hurried headlong, with no design for the ultimate structure. A man-made environment was the obvious goal – but who was the responsible architect? No one in my country. Not even the king of Great Britain or the president of America. Each inventor and producer who worked on building tomorrow's world just threw in a brick or a cogwheel wherever he cared to, and it was up to us of the next generation to find out what the result would be.

They taught us in school about the human brain. They taught us that it stopped growing at about the age of twelve. Yet now, at sixteen, we were all still treated as if we had only half a brain. How could adults believe that people of my age would think more clearly, once our freshly developed minds had been pushed through the education machine and filled to capacity with the doctrines of the elders? It was *now*, while we were still young, that we had to think; it was now that we had to hurry and judge, if we were not to be drugged into accepting blindly the seats offered to us on the engineless train of the elders.

In school, the topic of progress was dealt with in an indirectly dishonest way. We were taught to believe in progress from Paradise. Our teachers seemed to me to be walking tightropes, juggling with the Bible in one hand and Science in the other. They had succeeded to the extent that God, it appeared, had figured out how to create man. Darwin had discovered how he had done it, through evolution, making monkeys first. We were also told that God created the world, with all its living species, in six days, but the Bible also said that to God one day is a thousand years and a thousand years one day. Einstein agreed that time was relative; this far the adults agreed among themselves. Natural science had even come to the same conclusion as that written in Genesis thousands of years ago: life on earth began in the ocean and not on land. Only when the salt seas were swarming with life, and the air was filled with the winged species, did the day come when the creeping things and all the beasts of the earth began to move on dry land. The unknown sages who wrote the book of Genesis were even supported in their claim that man was the last of all to come into existence in a ready-made world of plants and beasts. Everything rotated and ticked, everything functioned, before man, through no effort of its own, received its lungs, its heart, its senses, and its brain, and began to multiply.

No errors had been committed in the complex design of Paradise. Even here, the Bible and Science agreed. God was completely satisfied with the world he had created. He found it very good, so perfect that on the seventh day he stopped his work and rested – abandoning man naked, but well provided for, like any bird or beast before him.

According to science, man could not have evolved from beasts if nature had not provided amply for his survival. So far, there was full agreement.

But whereas God was pleased with his job, man was not. God was sure he had given man a perfect environment, an earthly Paradise. Man did not agree. While God rested, man took over.

Man wanted progress. Again, progress from Paradise.

Men, too, worked six days, and thought they pleased God if they rested on the seventh, although men quarreled about whether they were to rest on Sundays, on Saturdays, or on Fridays. But whether Christians, Jews, or Muslims, they hurried back to work again on the eighth day and continued their struggle to make a better world. For centuries and millenniums.

God had not thought of inventing dynamite. Did he realize his own shortcomings when he saw what we could do? Did he approve our remodeling of almost everything he had done? Religious adults seemed to believe that God guided our brains to ensure that any step we took meant progress. Yet we were taught that God had left us responsible for our own planet, with ability and freedom to build and destroy, to advance and retreat, to rejoice and suffer, guided purely by the intelligence, intuition, and conscience we had been given. Surely, this had to be correct if it were true that the creator would reward or punish us in afterlife according to our behavior. What really bothered me was that adults said that God had created nature, yet they acted as if the devil was on their heels unless they continued to sever their ties with nature. Even atheists, who argued that nature itself had produced man, acted as if nature was man's old and innate enemy.

At about the age of sixteen, I started to feel uneasy. My confidence in adults began to be shaken. They were not smarter than us kids. They just had fixed ideas and stuck to them even if they disagreed. They were dragging us along a road to an unknown destination; they had no goal, only something to escape from: the natural. A terrible war had recently raged. Now they were inventing new types of arms, worse than ever before. Disagreements in politics, in morals, in philosophy, in religion. Who could feel safe in following in the footsteps of such a generation? It was better to begin to look for a safer side track. I began to feel like a prisoner calmly preparing to jump off a train that was on the wrong tracks.

This was about 1930. Nobody with respect for himself would revolt against parents or school. My interests in natural history increased. I began to see not only the beauty but also the superintelligence behind the build-up of the world man had inherited. I took to the forests, the mountains, and the open seashore whenever I could, and became skeptical toward the trend of a civilization designed to take man further from this environment. Were we doing something mad?

Finally, I had to share my growing unease with someone of my own age. One day, while the class was busy in the changing rooms after gymnastics, I sat quietly, deep in my own thoughts.

'I don't like machinery,' I blurted out to the boy next to me, who was struggling with his shirt.

'You don't say?' He merely grinned back with such an overbearing look that I felt like creeping into the shoe I was putting on and knotting the laces over my head. I had said something completely ridiculous. Not a word more came from my lips.

There was one boy in the class, however, to whom I gradually felt I could confide my secret. A huge fellow who did not care for sports like the rest, and who did not roam in the woods like me. He read books, wrote poetry, and liked to stroll deep into philosophical dreamlands. To Arnold Jacoby I gradually dared to open my mind. He listened with big eyes. I told him I was going to leave everything. *Every*thing. I was going to find nature. Somewhere in the tropics, where food could be picked from the trees. I was not going to spend my adult life in Europe, where disaster was lurking around the corner. Our twentieth-century Tower of Babel was either going to collapse or to lead man into another horrible, universal war. Better to stay far away. From now on, I had a confidant.

How could the dream be turned into reality? Careful preparations were needed. First of all, I had to build up my body and improve a rather shaky physique. There was hardship ahead. I had another friend, Erik, a huge, husky chap who for a while after junior high school had left us to go to sea. Life at sea for two years had given him the kind of muscles I needed. Erik, too, was skeptical about modern progress. He romanced about building an ideal community in the heart of Africa or in the unexplored plateau of the Mato Grosso in Brazil. My ambitions were not so big. I wanted only to find a girl who would share the experiment with me.

With Erik and his gang, I took for the first time to sport. Cross-country running in the forest, and skiing when snow fell. On our winter vacations, we started something scarcely known in Norway in those days: we pulled ski sledges behind us with tent and provisions, and slept in the wild mountains. Weeks away from people. Soon we left without tents, but with warm reindeer sleeping bags bought from Lapps in the far north, digging ourselves snug and sheltered caves in hard-packed snow, or slicing it into blocks to build igloos, Eskimo style. The cost of these trips was covered by articles I wrote, accompanied by photos and cartoons about our adventures. Our mountain expeditions got longer and wilder when I obtained a huge Siberian husky from a Norwegian sportsman, Martin Mehren, just back from crossing the then unknown center of Greenland with dog sledge and skis. With the dog, whom we named Kazan, pulling our food, Erik and I would build igloos to sleep in during the winter, sometimes even on the highest peaks and glaciers of Norway, in Rondane and Jotunheimen. Breathtaking views of the world below us at sunset or at moonrise showed through the door of our igloo. Only during such holidays in the wilderness did I really feel myself at peace. Better, under the blue sky, high above tree level, I felt literally on top of the world. During these holidays we were intimately exposed to all the elements, coming to grips with nature. These trips into the wilderness and the reading I did about primitive cultures in my mother's well-stocked library, had a greater appeal to me than the school textbooks. High school marks became medium, neither good nor bad, except in natural history. I did not care. I merely wanted to know how to get on friendlier terms with nature and how men and animals could once have thrived in the simpler environment of which we were all a product.

One desirable product was girls. I was desperately interested, but too shy to approach them. Since the time when my parents had forced me through three terms of dancing school as a boy, I had never dared to mingle with the other sex. They were magical, not real human beings, and I did not know how to talk to them intelligently. Yet I would never return to nature before choosing one of that enticing species for company.

It was at a graduation ball that I met Liv, at a restaurant on a pier in the fjord. Everyone was happy. Schooldays were over; everyone was dancing. Not me. I was sitting alone at an open window, watching the reflection of the moon tremble in the wake of small boats as they passed to and fro on the black waters. Suddenly, I found myself briefly in the company of a friend who parked an unknown girl from another town beside me. Bushy blond hair, laughing blue eyes. Sorry, don't dance. But what about a walk? No? Then let's chat. OK. Words and words. From jokes to philosophy. Damned intelligent eyes. Worth taking a chance.

'What do you think about turning back to nature?' I asked out of the blue.

That did it; she had grasped the very point.

That girl was to share in my experiment. Liv and I were to meet again as soon as we had both moved to Oslo, where we were to begin our university education. I would study, of my own choice, zoology and geography. She, pushed by her father, and to my horror, social economy. My choice of zoology as a main subject was obvious. Geography was to prepare myself for the experiment, to learn where it best might be conducted. My interest in aboriginal tribes and foreign cultures had not diminished.

Attention was now focused on Polynesia: on the Stone-Age people who had settled in the far-flung islands of the East Pacific. But if I were to study anthropology at any university, the course would devote only a few hours to all Polynesia. By good fortune, I was granted a better solution.

The world's largest private collection of books and papers on Polynesia happened to belong to a wealthy Norwegian wine merchant in Oslo. Bjarne Kroepelien had as a young man spent the happiest year of his life in the home of the great Polynesian chief Teriieroo on Tahiti, and on his return to Europe he began to collect anything published on Polynesia and the Polynesians, no matter where and when it was printed. Kroepelien was intrigued when informed of our secret plans. He let me use his important library as if I were a son in his house. Thus it happened that, whereas my formal training was in zoology, every spare moment, and more and more of my efforts, went into the world of books in Kroepelien's extraordinary library. (I did not know then that this library, after the death of Kroepelien, would be purchased by the Oslo University Library and become the backbone of the Kon-Tiki Museum research department.)

My animal studies never became quite what I had hoped for. We learned little of wild beasts and the way they lived in the wilderness. We sliced up intestines and looked at them under microscopes. We transplanted feet from the belly to the back of salamanders. We checked Mendel's law of inheritance by breeding thousands of small banana flies in bottles, and counting the inherited number of hairs on their backs. We went on excursions to haul in dragnets seething with the queerest wonders from the sea bed, but their life and function in the environment was ignored in favor of their Latin names. Was our university approach to nature superior to, or only different from, that of the Polynesians, who specialized in knowing animal behavior and in the way the existence of animals, dead or alive, could best serve man? I had to think like my fellow scientists now. And also as a

Polynesian, because one day in the future I was to live among them.

Liv moved to Oslo to start university one year later. She still wanted to share in my attempt to return to nature. But there were practical problems to solve, and year followed year as we trod on separate university steps, heavily laden with books on academic subjects. Mine on natural science, but hers on modern economy.

But how could we return to nature?

Only my father could help with a loan for a journey all the way to the tropics.

Only my mother could convince him to do so.

Only my university professors could make my mother feel that this was a sensible thing to do.

Only a scholarly project could encourage my professors to sponsor a field trip to Polynesia.

So, where to begin?

I had to work out with my professors an academic program that would qualify me for research in Polynesia. A solid education was needed, not only to get there, but also to come back capable of supporting a wife if, against all our expectations, we should be forced to return to civilization.

After seven terms and consultations with experts in Berlin, a project was developed and sponsored by my zoology professors, Kristine Bonnevie and Hjalmar Broch. I was to visit some isolated Pacific island group and study how the local animals had found their way there. How the fauna had developed on truly oceanic islands that had never been attached to the continents. Islands that have risen, sterile, from the bottom of the sea as smoking flows of molten lava. When the lava cooled, the various living creatures must have arrived by swimming, flying, drifting, being blown – or perhaps by getting a lift from human voyagers? Winds and ocean currents were obviously major factors in the transfer, and geography as a side topic was especially useful.

Thus it happened that, in the fourth year, my professors persuaded my mother, and my mother my father, to grant me a loan big enough for the necessary tickets. I wanted nothing to cover living costs at our destination, since, according to the fundamental nature of our project, there would be nothing for us to buy or rent.

As the barriers seemed to collapse around me, Liv's turn came; she was still mentally chained to her parents, who lived in a small town south of Larvik. So far, her sufferings had been restricted to the tender soles of her feet. (As contribution to a common cause, I had insisted that she take her shoes off whenever we had a Sunday hike in the forest, for our soft skin had to be prepared for the jungle floor.) But now the time had come for her to throw into the wastebasket all her own studies in economics, on which she had spent much of her own time and her father's money. We had to get her parents' consent. She was not yet of age. My own mother had approved of Liv. She was genuinely delighted at the idea of my taking with me a young woman she adored rather than going alone to a place famed for its hula-hula girls. My father's less romantic reasoning brought him to quite the opposite conclusion, just *because* of the hula girls. But he finally and reluctantly yielded to my mother's pressure and Liv's charm.

The hardest moment came when Liv had to write a letter to inform her own very respectable parents. Embalmed in a thousand fine phrases, the naked messages in her letter to her old-fashioned parents were:

No more economics.

No more civilization.

Marrying and leaving for the Marquesas Islands.

Horrified, Liv's mother had read the letter aloud. Slowly, Liv's father had raised his powerful body from the easy chair and headed for the bookshelf. The encyclopedia! M for Marquesas.

Good God! In the mid-Pacific! And the old text spelled out that the islands were renowned – for cannibalism and fornication. My future father-in-law clutched his heart and staggered back to his chair.

Fortunately, neither Liv nor I were present. Mental dynamite had been detonated, and it took time, more letters, and the calming interference of my own father, before the infuriated father-in-law-to-be felt at ease and agreed to let his only child be carried away to the Marquesas by a young stranger.

Liv was only twenty and I was twenty-two when we suddenly felt ourselves free from all ties.

The green light was everywhere. Nothing to stop us from launching the project of our dreams.

Farewell to civilization: Back to Nature.

III

Ticket to Paradise

FOR THE THOUSANDTH TIME, we pored over the colorful map of the South Seas. For the thousandth time, we sailed around on the vast ocean, our eyes scanning the blue paper, hoping to find a speck suitable for us. A single virgin speck among the thousands of islands and atolls. A speck that the world had overlooked. A tiny free port of refuge from the iron grip of civilization.

But every tempting little niche was already swept off the map with a penciled cross. It didn't suit us. This we had learned from heavy volumes of geographical literature.

Rarotonga crossed out. A motor road encircled the island.

Moorea crossed out. Hotels and tourists.

Motane crossed out. No drinking water.

Hututu crossed out. Barren of fruit trees.

This one crossed out, naval base, and that one crossed out, too small and overpopulated. Soon the map was speckled with small crosses. It began to resemble a celestial map in terms of practical use for us.

To live with bare hands like early man demands a lot from an environment. It would have to be fertile and luxuriant, and unclaimed by others. But wherever there was fertile land, it was densely inhabited. Wherever it was uninhabited, the environment was too poor to sustain man unaided by some degree of culture. So we had crossed out the large continents one by one. Bit by bit, land by land. Next the net was drawn around the South Sea Islands, tighter and tighter, cross after cross.

Just below the equator, where the mild trade winds flowed westward across the map like arrows, lay the thirteen islands of the Marquesas group. They had already become thirteen crosses. But we returned to these alluring islands with an eraser after the entire map had been filled with crosses and every single island rejected.

Nuku-Hiva, Hivaoa, and Fatu-Hiva were the largest. Fatu-Hiva, the most beautiful and verdant island in the South Seas. Over and over again, we brought out the few pictures and the sparse information available on Fatu-Hiva.

Once upon a time, 100,000 Polynesians were supposed to have lived in the Marquesas group. Today, a mere 2,000 were left, with only a handful of white men. The Polynesian islanders were dying out at a tremendous rate.

And Fatu-Hiva was the most luxuriant island in the South Seas!

If 98,000 had disappeared there had to be enough room for two of us. There had to be a peaceful spot among the abandoned ruins. A spot remote from all diseases. A spot where civilization had not taken root, but where fruit abounded in overgrown, abandoned gardens.

Maybe we could find a deserted valley, a lonely plateau, or a small, fertile bay. There we could build a home. From sticks and leaves. There we could toil for our living in the forest. Live on fruit and fish and eggs. At one with nature, among palms and foliage, among birds and game, in sun and rain.

There we could make our experiment. Go back to the forest. Abandon modern times. The culture. The civilization. Leap thousands of years into the past. To the way of life of early man. To life itself in its fullest and simplest form.

Would it work? Yes, in theory. Theory didn't interest us. We wanted to experience it. We wanted to see if the two of us, man and woman, could resume the kind of life abandoned by our ancestors. If we could tear ourselves away from our artificial life, completely and utterly. To be independent. Independent of the least aid of civilization. Independent of everything except munificent nature.

So, the island of Fatu-Hiva became our choice. Mountainous and lonely. Rich in sunshine, fruit, and drinking water. Few natives and fewer white men. We drew a firm ring around Fatu-Hiva.

Outside, the winter fog was stealing in upon the city.

Thus it happened that, in a biting wind on a Christmas morning, we left on our honeymoon.

Tickets could only take us part of the way. To the travel agencies we were heading for a destination outside the rim of our own planet. Tahiti was the end of the world.

Three days' train ride down through Europe. From winter darkness, with snowflakes dancing and rice hurled at the train window by Arnold and former classmates, to the bright Mediterranean light blinding us as we rolled up the curtain on the express rumbling into Marseille.

Once a month a French steamship left this port for Tahiti. A single Norwegian cargo ship with the same destination sailed at irregular intervals from San Francisco. There was no other way to reach Tahiti. And the sea voyage took six weeks. We saw our first orange trees in Algiers and coconut palms in the French West Indies before we crossed the Panama Canal and the big ship began rolling smoothly into the waters of our dreams: the Pacific Ocean.

In time, Tahiti rose above the horizon. We could smell a tropical aroma, the warm, spicy fragrance of tropical greenery, before we saw the hazy blue peaks of jagged mountains above the western horizon. Six weeks with nothing but the familiar smell of steam engines and salt sea made passengers from nineteen countries line the railing, sniffing the mild air, and trying to see land below the peaks.

The large island now rose from the sea as if dripping wet, with white surf pouring from its coral reef. Mountains wilder than sharks' teeth bit into the trade-wind clouds of the blue sky. The famous Diadem and the mountain Orohena soared among green hills seven thousand feet above the palm-lined beach. Gauguin, Melville, and Hollywood had not exaggerated. Nature itself had exaggerated. Something so beautiful seemed impossible.

We began to sing Tahiti's national hymn, tender and peaceful, composed by some former island king: '*E maururu a vau*' (Happy am I). Poets and painters, businessmen and colonial clerks, tourists and adventurers, we all felt that we were approaching our lost Garden of Eden, lost in the middle of the sea. Here it came, unspoiled, toward us, sailing over the sea, verdant as a huge flower basket.

Finally, we heard the surf. Soon we saw a red church spire piercing the compact roof of tropical foliage. More houses. Papeete, the capital of French Oceania. The engine slowed down. We slid through an opening in the coral reef, the surf frothing around us. A calm harbor. A huge warehouse with a metal roof and a wharf packed with people. No one was wearing a loincloth. All were dressed like us. An exhilarating atmosphere of jubilation and song among the multinational passengers on board. We wanted to show them that we were an informal breed of mankind, uncomplicated and free, like them. Ashore a few hats were lifted and a few hands waved back from the unperturbed crowd, while customs and immigration officers in white uniforms saluted army fashion as they came up the gangway. A sweet, too sweet, smell of copra from the warehouse filled the warm air. Thousands of sacks were waiting for the ship. This was what it had come for.

About 20,000 people lived on Tahiti at that time, mostly of pure or mixed Polynesian blood. Papeete was dominated by Chinese merchants, who owned all the little stores and a couple of tiny restaurants, in addition to countless carts selling sweets and other merchandise in the streets. The governor's palace, a post office, a handful of French stores and colonial offices, a bamboo cinema, churches, two primitive hotels, and a few rows of wooden bungalows made up the rest of the town. Beyond, forests and fern-clad hills quickly rose toward the central spires of the Diadem, and on both sides of the town a narrow belt of flat lowland, covered with coconut palms, banana plants, giant breadfruit trees, citrus fruits, mango, and papaya, extended in a ring all around the island, with native homes scattered between the trees. We were longing to get into this exotic world, away from the town.

After a couple of nights in a local hotel, where neighbors on each side could look down on us by standing in their beds and peeping over the partition walls, we decided to take off for a brief visit to the country. No schooner was yet scheduled for the voyage to the Marquesas.

An open bus, with more people, pigs, chickens, and bananas hanging outside or on the roof than were piled inside on the wooden benches, bumped along on a dirt road following the calm lagoon surrounding part of the island. On each side of the town, this road ended in the bush; there was not yet a road all around

Tahiti. Ten miles eastward, on the north coast, was the valley of Papenoo, running steeply down from fern-clad mountains to the level coast, where a river emptied into the sea.

The protective coral reef was broken here, and ocean swells thundered against the rattling pebbles. A lush, wild garden. Tahiti at its best. Here was the home of the supreme chief of Tahiti's seventeen chiefs, Teriieroo a Teriierooiterai, Kroepelien's friend. This was where the spark had been struck that started the library I knew so well. This was where Kroepelien had met Tuimata, Teriieroo's daughter. He himself had buried her when thousands of healthy Tahitians had died of the Spanish flu that swept the island after World War I. Kroepelien had helped to carry away wagonloads of dead islanders. His own book on Tahiti ended at Tuimata's tomb. There, he wrote, his own heart lay buried. He never wanted to go back to the island. But he had sent with us a parcel of personal gifts for his old friend, Teriieroo.

Chief Teriieroo met us at the steps of his bungalow, huge and friendly, strong and well fed. His new wife, who clearly provided well for both of them, followed two paces behind with an equally hospitable smile. Both came barefoot and each had a brightly colored pareu wrapped around the body, he from the waist down, she from above the breasts. We felt welcome already as they approached us, before they knew our mission.

Bjarne Kroepelien! *Bjarne!* Were we friends of Bjarne? There was no end to the rejoicing. We were almost carried up the stairs, and it was no longer possible to get back to the bus or Papeete. We were literally kidnapped in the Papenoo Valley. This was to be our first home together. Teriieroo's friends in Papeete would send word by the bus when some schooner captain decided to set sail for the Marquesas. Perhaps next month. Perhaps later. Nobody went to the backward Marquesas except when a schooner captain felt he could profit by fetching a boatload of copra.

We came to look upon Teriieroo as a sort of island king, noble and righteous, and thus he was generally regarded by the other islanders, although all real power was delegated through the French governor in Papeete. Teriieroo had won the Legion of Honor for his loyalty to France, but he had no personal ambitions beyond bringing joy and justice to his own surroundings and to friends of friends. A not unimportant procedure was to invite guests and entertain them with the most sumptuous Polynesian repasts, affording equal pleasure to the palate and to the mind. Everything served was picked or caught by Teriieroo's own hands and those of his sons and made into palatable, flower-ornamented dishes by his wife, Faufau Taahitue.

In Papenoo, we had a genuine introduction to the Polynesian way of life. There was no idleness, no boredom, no rush, and no waste. The soil, the sea, and the river provided what the little community needed, and no effort was made to exhaust the resources in exchange for wealth. Wealth in Papenoo was not measured, as among us, by counting what one had; what counted there was how one felt. In my new friends, I recalled what I had read; in Polynesia, both pleasure and prestige were obtained by those who gave away the maximum of material possessions. Polynesians were loosely attached to personal property.

Teriieroo was not only a great personality but also an outstanding speaker. During feasts, he showed that he was a gifted orator in both French and Polynesian. For me he became a new teacher, a specialist in all the practical details I had not learned from books. A pureblooded Polynesian of the clearly vanishing type, he was one of the few I was ever to meet who took pride in his ancestral culture. He was not convinced that the European way of life had brought only blessings to his island. Our own plan to try man's original way of life intrigued the middle-aged chief. He told his wife he would have joined us on our voyage to the Marquesas had he been a few years younger. The islands up there were reputed to be something special. He had a friend who had been there. Up to a hundred coconuts on a single palm. Wild fruits abounded in the valleys. Particularly on Fatu-Hiva, the island farthest south. Oranges grew in the forests. Even Teriieroo's favorite food, *fei*, the red mountain banana, grew in the valley bottoms. Here in Tahiti, mountain climbers, their big toes set far apart like those of monkeys, had to fetch the *fei* from almost inaccessible precipices in order to sell them at the market in Papeete. And the brown rat from Europe had not reached the Marquesas, so there was no need to nail sheets of tin around the palm trunks to stop these rascals from stealing the coconuts. The chief was convinced that up there life was still as it used to be in Tahiti, where even *fei* and other rare bananas had formerly grown in the valleys. Today, he said, they die from imported pests as soon as anyone tries to plant them.

At meals, Teriieroo and Faufau were taciturn, as were those of his children who shared them with us. It was good manners to enjoy the food and not distract others by talking. A belch at the end of the meal was healthy and even a gentle way of telling the hosts that the food had been delicious. Talking began afterward. The first day, there were forks and spoons in front of us, and we all used them. But when the chief heard of our plans, he swept all metal off the table and showed us his fingertips. Clean. Good etiquette involved washing our hands just as we sat down to a meal. With the tips of two fingers and a thumb he broke off a piece of baked breadfruit and dipped it into a thick white coconut sauce, sucking the mixture deep into his palate and rolling his tongue. This is how to enjoy good flavor, he explained. You people, he argued, are so used to putting cutlery into your mouths that you do not realize that the metal upsets the flavor. Soon we all sat with the three permissible fingertips in our food and began to think that putting metal into our mouths was barbaric.

My young bride sweated with sticks and branches over hot stones in Faufau's earth oven, learning how to make Polynesian roots and fruits edible and tasty. The chief took me upstream or along the seashore to gather raw material. There were prawns to be caught with bamboo traps in the mountain stream. In the lagoon a vast variety of fish and crustaceans, octopus and mollusks of various kinds could be caught with net, hook, spear, or with our bare hands. Edible roots were identified by their visible leaves. Not everything that looked like food was edible for man. Some fish, roots, and fruits were poisonous. Shark flesh would become perfectly good if sliced and left overnight in water. Seafood did not need to be cooked; it was enough to cut fish into cubes and leave them to soak overnight in lime juice. The red mountain banana could not be eaten raw nor could breadfruit unless buried in the ground until entirely fermented. Manioc was a dangerous root if not grated and its poison filtered out. The best wood for making fire by rubbing was the bone-dry branches of *borao*, the hibiscus tree, split along the pith.

In Teriieroo's opinion, we could safely abandon all the tools of civilization, but there were two utensils even he could not do without: a cooking pot and a long machete. Without the pot, too much of the jungle food would be indigestible for modern man. And without the knife we could not even sharpen the crowbar needed to get the crashproof outer husk twisted off from the shell of the coconut.

On the boulders in the river, there were snails with spines on their shells, looking almost like tiny sea urchins. We had learned that they were painful to step on, especially with bare European feet that did not have the leatherlike soles that had developed on our barefoot friends. I watched my steps carefully one morning as I waded in the river looking for prawns. Perhaps there would be more if I got across to the other bank. The water was deep in mid-stream, and swirling, and it was impossible to see the bottom. It was then that I stepped heavily on one of the damned snails, and lost my balance. There was too much current to get another foot-hold, and I was swept headlong with the whirling water toward the estuary. This would have been nothing to the average swimmer; it just happened that I was not one of them.

My greatest shame as I came of age was that I could not swim at all. As a child I had twice been a hairbreadth from drowning. One winter at the age of five I had tried to jump along with the big boys, who knew the trick of landing with both feet on a floating block of ice and then leaping back on solid ice before the block capsized. I was not fast enough, and rolled around with the big ice cube, ending under the main ice sheet covering the lake. Seen from above, the ice is white and the open hole looks black. But seen from below the hole is bright and the ice cover is dark. I fought desperately to come up where I thought the dark open water was, only to bump head and body against a compact cover of ice. More dead than alive, I was pulled up legs first by the big boys.

Some years later I was playing tag and tumbled off a high bridge into a surging sea. Struggling desperately, I swallowed a lot of water. I was sinking helplessly for the third and final time when a life buoy was hurled from the bridge and saved me from the deadly grasp of the deep water that I came to fear and shun for years. No one could convince me that by moving my limbs I could float.

Now I was underwater again, rotating, gasping, and waving, carried in the rushing rapids like a sack of potatoes toward the roaring ocean surf. The waves collapsed in tumbling walls as if pushed against the boulder beach by a thousand tanks. A deafening cannonade, thundering seas, a roaring choir of rattling boulders. In seconds, I would be smashed to bits. Quickly. Panic must yield to sober resolution. Self-conquest. Steady. With long calm strokes, I began to move. I knew how to, but had never tried. With the greatest ease, I was free from the rapids and swam to the river bank, the boulders resounding like gnashing teeth in the foam from the deafening breakers a stone's throw beyond me. I stood for a long time watching the ocean fury from which I had just been spared. I felt a great triumph. I was never going to be afraid of water again. The tropical sun was hot. I went higher up, heading for a deep, calm place in the river. I plunged in and began to swim like a frog. Teriieroo joined me. I did not tell him I had never swum before.

Teriieroo was too heavy to show me how to walk up a coconut palm. His grandson Biarne, named after his friend, put his feet flat against the trunk, shot his body out in an arch and wriggled up monkeylike on all fours, as easily as I would have crawled along the floor. But I knew how to climb a branchless pine, and I embraced the palm trunk with arms and legs, climbing up in good northern fashion, chest against the trunk. Triumphantly I found that the coconut palm was easier to climb than any pines, for the surface was indented with shallow rings around the trunk all the way up to the huge, fernlike crown. Up there I proudly waved down to my friends and tried to tear loose a coconut. But I could not. My breath was gone. Time to get down. I tried, but could not. The rings that had helped me up now prevented me from sliding down. The almost imperceptible edges were all directed skyward. There I hung, clinging to a branchless trunk. This was when Teriieroo shouted a warning. One of the coconuts was not a nut at all, but a big wasps' nest! The inhabitants began to swarm forth, buzzing unpleasantly. I tried to arch my body out to get down Polynesian style, but almost fell. I clung to the palm again, and let gravity work. The pain of an extremely sudden return to the soil was overshadowed by the feeling of parting from a tree that had retained a considerable part of my skin. I felt as if my behind had been worked over by a hammer, and my front by file and sandpaper. Teriieroo discovered that half my big toenail was loose and put his 275 pounds behind the job of pulling it all off. Two weeks passed before I learned to climb a low palm and twist off the tough stalk attaching the coconut to the tree, carefully watching out for poisonous centipedes and wasps' nests.

Four weeks passed in Teriieroo's home before the bus brought a message that Captain Brander of the copra schooner *Tereora* was planning to set sail for the distant Marquesas.

Before our departure, a special party was prepared on the floor of Teriieroo's big balcony. A long carpet of fresh green banana leaves was laid out as a tablecloth and studded with aromatic flowers. Colorful leis, garlands of fern leaves, and sweet white *tiare* flowers produced an atmosphere of gaiety and happiness. The juicy vapor of baked bush pig and chicken rose from a stone-lined earth oven when Teriieroo and I returned with a basket full of jumping prawns. Women had been fishing in the lagoon and pulled up taro and sweet potatoes. Children and children's children had been shaking branches or climbing big jungle trees for oranges, papaya, mango, and breadfruit. Nothing pleasant to the palate should be missing that special evening.

No banquet, however professional and extravagant, can give the guests more pleasure than a genuine Polynesian *umu* with juicy-fresh delicacies served straight out of a steaming underground earth oven and consumed under the stars, the nostrils filled with sweet fragrance from tropical flowers, the ears with soft drumbeats from the distant reef. Culinary art has always played a key role in Polynesian culture. Imitation South-Sea-style meals served in Continental restaurants can be compared to Polynesians trying to play Beethoven with bamboo flutes and sharkskin drums.

But on that evening Teriieroo broke his own rule. He gave a speech while we all had our fingers in the gorgeous food. Big, flower-ornamented, and comfortably wrapped in his favorite pareu, he got to his feet at one end of the green carpet and pointed to Liv and me, who were squatting Polynesian style at the other end. First in Tahitian and then in French, he told his many guests that, as they all knew, he had twenty-nine children, but now he was going to adopt two more. In doing so, he also had to give them new names, since the old ones were Norwegian and too much of a tongue twister for a Tahitian. Could anyone pronounce Liv or Thor? They all tried, one by one. 'Rivi' and 'Turi.' Wild amusement. Nobody could.

That is why Teriieroo and Faufau adopted us under new names everybody could immediately repeat, except us. The chief had named us Teraimateatatane and Teraimateatavahine. It took us all evening, and contributed vastly to the enjoyment, before we learned to separate the components of our new names and pronounce them properly: Terai Mateata Tane and Terai Mateata Vahine.

Mr. and Mrs. Blue Sky.

Only now were we ready for our real introduction to Polynesia.

IV

Cutting all Ties to Civilization

Tahiti lay a thousand miles behind us when the islands we had dreamed of rose from the sea one morning with the sun. We had been rolling and pitching for three weeks, sleeping cramped in between Polynesians with pigs and chickens on the deck of a little copra schooner.

As the sun rose glowing red on the eastern horizon, our first Marquesan Island seemed pale blue, like shadows of fingers, on the northern horizon in front of the dancing bow. Steep, rugged, and menacing, the mountain masses hurled themselves ever higher as we sailed on, until they were soaring rock fortresses high above the surf. Tumbling, frothing, and rumbling like a distant thunderstorm, the endless sea beat wildly against these fixed obstacles in a world of living water. From a distance, the islands seemed far from hospitable. Uapou's incredible pinnacles emerged from the water like shadows of reversed icicles, but as we came nearer their color changed to a warm jungle green. As the schooner stood in still closer, we seemed to be approaching the ruins of a seagirt castle, with wisps of cloud sailing around the towers like smoke. Then the palm beaches also rose, and we coasted alongside a lofty island, smaller but wilder and more spectacularly beautiful even than Tahiti.

As one island rose from the sea, another sank and disappeared, for there was a great distance between the islands in the Marquesas group. The Pacific stretched between them in many tints of blue, but around each island the water was as green as grass, due to masses of microscopic plant plankton thriving upon an incessant rain of minerals gnawed from the brittle rocks by the perpetual surf. Shoals of fish were attracted to this ever-green marine pasture, with dolphins and birds in visible pursuit. Swarms of seabirds followed the little schooner and plunged after the fish that continually struggled on the line we were towing astern.

We were much nearer the equator now, and as we came inshore, we could verify that here the Pacific reached its highest degree of fertility. One valley after the other opened in front of us and closed behind us as we sailed past, all formed as deep, wild gorges cutting into the central mass of ridges and peaks. Only truly vertical precipices had managed to shake off the jungle and rise as naked red rock above the chaos of luxuriant greenery that flowed down the steep ridges and bluffs to the palm-studded valley bottom.

The tropical heat alone was not responsible for this extravagance. In the interior of the islands, the towering peaks intercept the westward course of the sparse but ever-present little trade-wind clouds, and squeeze the rain out of them before they manage to proceed westward. Fresh rainwater always pours from the mountains in torrents and rivers, through dark jungles and friendly valleys, down into the green sea. Everywhere the tooth of time had gnawed greedily into the fragile volcanic rock. Caves and subterranean streams, pinnacles and grotesque carvings in the mountains, turned the whole scenery into a fantastic fairyland. In this exotic environment, we were to be set ashore. Here, somewhere, we were to dive into the unfamiliar jungle while the *Tereora* returned to the twentieth-century world. Once left behind on Fatu-Hiva, we should be deprived of telegraphic contact, in fact, of any contact whatsoever, with the outside world ... no message from anywhere until some unscheduled schooner should happen to call again at least several months in the future. If a war broke out, nobody could inform us. We had learned in Tahiti that nobody in the Marquesas Islands had heard that World War I had broken out until it had ended.

The Marquesas Islands lay closer to the equator than any other island groups in Polynesia. On our way northward from Tahiti the schooner had first zigzagged through part of the Tuamotu archipelago and called at the low coral atolls of Takaroa and Takapoto. There we had seen how the civilization we were trying to avoid was slowly radiating from Papeete into surrounding Oceania. The trading schooner was the bringer of culture and was a profitable business enterprise. It carried a well-stocked store below deck, and by selling at high prices, doubled its business by getting back with a profit the same money paid to the islanders in return for bringing the heavy loads of copra on board.

'It's all crazy,' said Captain Brander of the *Tereora*, a jovial Englishman with white hair and a red nose, who loved the islands and his whisky, although he never set foot ashore. A sort of retiring island Santa Claus. We had learned in Tahiti that this Pacific old-timer was a college graduate who had wanted to escape from it all. He admitted it.

'Crazy. But they want it, like everybody else. I detest our own civilization; that's why I am here. Yet I spread it from island to island. They want it, once they have had a little taste of it. Nobody can save them from the avalanche. I certainly can't. Why do they want sewing machines and tricycles or underclothing and canned salmon? They don't need any of it. But they want to tell their neighbors: look here, I've got a chair while you are squatting on the floor. And then the neighbor also has to buy a chair, and something else not possessed by the first one. The needs increase. The expenditure. Then they have to work, although they hate it. They harvest coconuts and split them to dry the kernel into copra to earn money they absolutely don't need.'

As usual, old Brander had remained on board as the *Tereora* anchored in the lagoons of two of the Tuamotu atolls a good three hundred miles north of Tahiti. Invariably it was his trading master, the brilliant Tahitian supercargo Théodore, who went ashore to attend to business. Brander only brought the schooner to the desired destination. With Théodore, we had climbed into the lifeboat and waded ashore on the low coral atolls to see what was going on. The natives were unloading corrugated iron and window glass in the baking sun. Others were wading out again with heavy sacks of copra. As we were invited into the home of an islander to get some shade from the midday sun, salesman Théodore triumphantly pointed out an old, discarded iron stove standing on the floor. He had sold it. There was no stovepipe attached, nor any chimney on the roof, for the tropical climate did not require that the oven should ever be lighted. The rusty thing was set up as a piece of costly European furniture.

On the second atoll, we had hardly set foot ashore when a group of excited Polynesians pulled us along between the palm trunks. There, in an open place of hard coral sand, stood a tall old automobile, motionless among the palms like a long-legged foal straddling to keep upright. Flat tires. No road. For a modest fee, we were pushed onto this rolling royal throne, the pride of the whole island, and its envied owner actually managed to wind it up for a painful shake, dancing us around half a dozen trees back to its original site, accompanied by a host of pedestrians.

Nothing similar seemed to have reached the Marquesas group.

We were first landed on the main island, Nuku-Hiva, in the northernmost part of this widely spread archipelago. Here was the residence of the French administrator, who was also the only doctor in the entire group. He had no local means of inter-island communication, since the lack of any sheltered harbor made it impossible to have larger boats than could be carried up on land. The Marquesas Islands have many open beaches of boulders or black volcanic sand, but no deep bays. The entire shoreline rises so abruptly from the bottom of the deep Pacific that no coral polyps have managed to build a protective reef around the shore as at Tahiti.

Before we left Europe, we had needed a special permit from the French colonial office to go to the Marquesas. French law forbade any visitor to remain ashore there for more than twenty-four hours. Was it to protect the island? Or was it to protect the visitor? Nobody had been able to tell us the reason.

From Nuku-Hiva, the *Tereora* turned south to visit the other islands in the group before the home voyage to Tahiti. The schooner went under sail, but had an auxiliary engine in case the wind failed. We slept in double rows on the cabin roof above deck, as the deck itself was awash in rough seas. We slept in a row with a common rope over our chests and under our arms in case of excessive rolling.

In the daytime, we lay on our stomachs trying to penetrate the deep jungle with binoculars as the island coasts slid by. We were curious about the least detail, for one of these islands was to be our home. Beautiful, spectacular, but heavy, sometimes almost gloomy and frightening.

Brander watched us as we lay literally bewitched, gazing toward the promised land. We felt small in this tremendous landscape, yet, we were drawn to it as if by a spell.

'Austere and oppressive,' said Captain Brander. 'The heavy jungle and mountains seem to squeeze you small.' He wanted us to come back with him to Tahiti, but we refused.

Our next call was Hivaoa, the second of the two major islands in the group, and our last stop before Fatu-Hiva. Brander strongly recommended that we get off there. This was the last spot from where we could contact the outside world. There was a little one-man radio station, a French gendarme, an English shopkeeper, and a Tahitian male nurse. One more European, a Norwegian copra planter, lived in a valley on the other side of the island. Paul Gauguin had spent his last years there, and we were shown his lonely tomb. But nothing doing. We wanted the island we had circled on the map.

Next morning, we awoke in sheltered water as we slid in under the lee side of Fatu-Hiva's high mountain ridges. Shaped like a bean, this island is divided lengthwise from north to south by a knife-sharp comb, its two main valleys opening toward the sheltered west coast.

'Where do you want to be let off?' Brander muttered as we came close to the rocks of the northern cape. He admitted he did not know this island, and his map could not help us. The old sailing chart showed no details except the coastal contours and the available anchorages off the two main bays. It was decided that we should follow the coast as close as would be safe, and pick the place we wanted.

Sails went down, and with the engine at slow speed, we went closer inshore. Sheer, naked cliffs seemed to be hanging above our heads, plunging straight into the splashing surf. But as we moved along, the mountain curtains seemed to be drawn aside, and one by one truly glorious valleys opened to view, curving on to be lost in the island's interior. Out at sea, the jungle air met us again, stronger than ever. Fatu-Hiva. A rock-walled greenhouse.

Brander and Théodore for once were hanging on the rail as we were, speaking in Polynesian to one of the crew, an islander who seemed to know Fatu-Hiva well. The question was for us to locate a place where we could be set ashore with the best chance of subsistence. The first requirement was drinking water. As we spoke, a mighty valley opened before us. It looked completely artificial, like the stage of a theater, with rows of red side screens jutting into the green palm forest from both sides. These fantastic side curtains were outlined with bizarre profiles against the greenery, as if cut from plywood by an artist with a sense of shape and effect, rather than crumbling red tuff molded by millennia of rain and storms. A row of thatched bamboo sheds was discernible between the palm trunks above the boulder-strewn beach. Shelters for canoes.

'Hanavave,' Brander explained, and nodded toward land. 'Here there is plenty of water in the river and an abundance of fruit all the way up the valley. About fifty natives live here, according to this fellow. All in one village.'

Liv was fascinated and immediately wanted to get ashore, but Brander shook his head.

'An unhealthy climate,' he said. 'The valley is very moist and the air filled with vapor. The natives here suffer from all kinds of diseases that may infect you too. Elephantiasis is terrible here in Hanavave.'

With silent awe, we saw the mighty, theatrical rock curtains of Hanavave being drawn as we passed by. We were never to see a more beautiful composition of natural scenery. At short intervals, narrow gorges and ravines opened in passing review, filled with luxuriant jungle vegetation. But they were too small to ensure enough wild fruit for our survival. A friendly beach came into sight, with a palm-studded fruit forest crowding right down to the dark sand.

'*Aoe te vai*,' explained our Polynesian cicerone. No drinking water.

One little valley after the other passed, each severed from the next by powerful walls and precipices. Either they lacked water or they were deep and dark like canyons, too austere for us to live in.

Still another pretty little valley, with a waterfall. Still another hope. But here in Hanaui lived an old native couple with legs like hippopotamuses. We lacked the courage to go ashore, even though we knew that the filaria of elephantiasis were spread by the mosquitoes and not by human contact.

The last valley before the southern cape was Omoa. Wider and more open than any other we had seen on Fatu-Hiva, it disappeared in a big arch into the heart of the island. Exhilarating, picturesque, although not quite competing with Hanavave, the Omoa Valley abounded in wild fruits and drinking water. We could see with our binoculars how a river poured its frothing wealth of water over a boulder barrier into the bay, where the *Tereora* was now slowing down. Our informant estimated that a hundred natives lived packed together in a village close to this bay. Farther in, the large valley was entirely uninhabited.

'Here or nowhere,' said Brander as the anchor rattled to the bottom. 'You have seen your choice. Want to return with me?'

'We have only seen the leeward of the island,' I insisted, and pointed to the long inland ridge that rose like the jagged crest of a dragon all along the axis of the island. 'What about the east coast?'

Brander and Théodore, supported by the well-informed islander, were quick to explain that no schooner ever passed along the other coast. There was no anchorage and not even a safe place to land a lifeboat. The last to try was smashed, for the large ocean swells break with unimpeded force along that coast, coming westward with nothing to stop them all the way from South America, four thousand miles distant. The east coast was in fact barren and deserted. The tribes once living there were extinct, and nobody even tried to harvest the copra on the other side of the mountain ridge.

I took a last look at the inviting green valley and the sky-scraping crest behind it. If we wanted to reach the east coast, maybe we could climb across.

We agreed to be set ashore in Omoa. The lifeboat was lowered and we said farewell to everybody on board. Brander tried to persuade us for one last time; then he said he would be sure to pick us up on his next call, perhaps in a few months, perhaps in a year. The swells were rolling gently under us in the boat. They came into the open bay as backwash from the wild surf we could see and hear as it tumbled over the black lava rocks of the south cape. Strong native oarsmen laid their combined strength behind the oars, and soon we rode in on the high surf to a slippery boulder beach. Four Polynesians jumped overboard to hold the boat, while four others helped us wade ashore with our belongings. Surf and undertow threatened to capsize the boat and, before we quite realized what had happened, all the eight men were on their four thwarts, rowing with long strokes back to the *Tereora*.

V

Back to Nature on a South Sea Island

When we really woke up to what was going on, when years of dreams became reality in a matter of seconds, we found ourselves standing alone on an unknown boulder beach; the sails of the *Tereora* were hoisted, and we watched the two-masted schooner slowly moving away. We followed it until the white vessel dwindled and was lost among thousands of whitecaps that filled the ocean.

There we were on the beach, with our luggage beside us on the boulders. Two big suitcases containing Liv's wedding gown, my formal suit, and all the usual apparel we had needed on the long journey as first-class honeymoon passengers from Norway – none of it was useful now. A couple more cases containing bottles, tubes, and chemicals for collecting zoological specimens – nothing to eat. I looked up into the palms fringing the beach.

Coconuts. They gave me back some tottering courage. We should not starve. I drew a deep breath and looked at Liv. We both laughed humbly and reached for the suitcases. We had to go somewhere.

The sun and the singing tropical birds warmed us up. The sandy turf above the beach abounded in fragrant flowers, and a feeling of high adventure and happiness overtook us anew. Then ... we suddenly noticed people standing among the trees. There were many of them. Watching us. Nobody moved and nobody greeted us. Some were in loincloths and some in tattered rags of European make. Copper-colored to brown, all were varieties of the Polynesian stock. Most faces looked more cruel than those of their friendly relatives in Tahiti and the Tuamotu atolls. But a couple of the younger women and most of the children were beautiful.

Seeing that we hesitated, an old crone was the first to go into action. She shouted a few words that sounded like a flow of vowels, softer than the Tahitian dialect. I did not get a word. Just shrugged my shoulders and laughed. That made the old wrinkled crone bend over and shake with laughter. Others laughed with her. She ventured forward, followed by the rest, and, to my surprise and fear, she headed for Liv rather than for me. She licked her thin finger and rubbed it against Liv's cheek, who in her surprise was incapable of speech. The old woman scrutinized her own finger and nodded with an approving smile. Only later were we to learn that the spectators had trusted me for what I seemed to be, but they were sure that Liv was a Tahitian girl dressed up and whitewashed. The old woman did not believe there were women in Europe. The vessels anchoring on twenty-four-hour visits to the islands had brought ashore many white men who came for brown girls, but no white woman had ever come for a brown man.

In the moment of general excitement, we found that all our luggage was gone. We could not ask for it, nor did we depend on it, so we just followed the crowd between the palm trunks to an open place with an enormous banyan tree surrounded by a boulder bench. Scattered native huts were seen around, the most noticeable being a wooden shack with the dreadful corrugated-iron roof we now profoundly hated. Here we were met by a young and timid European-looking man who, as it turned out, knew French. We spotted our luggage on the wooden floor of his little bungalow as he invited us in, with the silent reception committee crowding up to the open door. Our very gentle host was far from talkative in either French or Polynesian, but to satisfy our curiosity he told us that his name was Willy Grelet, the only one on the island who had grown up with a European father. His late father was a Swiss who had married a local island girl and whose only friend had been Paul Gauguin, whom he hardly ever saw, since they lived on different islands. Willy seemed introverted and lonely, clearly keeping aloof from the rest of the village people, who obviously respected and admired him. We were to learn that he was a very honest person, even though he loved money, which he gathered wherever he could get it, and he was rich, but saved his earnings as there was nowhere to spend them. Between the poles supporting the floor of his bungalow, he had a primitive kind of shop, where the other islanders came to obtain merchandise in exchange for their copra, all of which Willy Grelet, as far as we could see, had under his control. The modest store contained matches, shirts, flour, rice, and sugar, as long as the stocks lasted. Little else could be detected in his bungalow, and the shop was open only at sunset, when Willy returned from his own copra work. Apart from this, his remaining passion proved to be hunting the formerly domestic animals that now roamed wild in the jungle; he therefore knew the island better than anyone else.

Before the sun set, it had so heated the iron roof that we could not sleep. The brown spectators kept pressing around the open door and sealed glass windows, and Willy seemed to be in no hurry to go to bed. The three of us remained seated around his kerosene lamp till the small hours, discussing our plans. Our island host clearly looked upon us as the strangest creatures that had ever come ashore, but he understood our plan. Far up the valley, we should find what we wanted, in the interior of the island, where the natives rarely went and where abandoned gardens were engulfed by jungle growth.

Most of the night we spent compiling a small dictionary. Laughter had helped us ashore, but we also needed to know a few words that would enable us to go further. I had prepared in advance a long list of key words in Norwegian, and now Willy helped us to translate them by way of French into Polynesian. The difference from the Tahitian dialect was noticeable in the first phrase we had to learn. In Tahiti, 'good day' was *ia ora na*: here it was *kaoha*. Consonants were not in great demand on Fatu-Hiva, and we struggled hard to distinguish between certain words:

no	=	*aoe*
I	=	*oao*
you	=	*oe*
he	=	*oia*
they	=	*aua*
two	=	*eua*
who	=	*oai*
rain	=	*eúa*

In addition, Oua and Ouia were local place names. The word for 'nice' was *panhakanahau*, while 'bad' was *aoehakanahau*. Nevertheless, the term they used for the 'Polynesian' language was literally 'human' language, a survival from the days when their ancestors thought that the white visitors were gods.

Polynesian words buzzed around my head with the mosquitoes as we crept to bed under Willy's large insect net.

From the beach we heard rhythmic thunder: the surf reminded us that we were on a lonely South Sea island, thousands of miles from anywhere.

Adam and Eve, when God drove them out of the Garden of Eden, must have had feelings the opposite of ours as we started our walk into the lush valley of Omoa at sunrise next morning. They left; we were arriving. The song of tropical birds resounded from all parts of the valley as they joined the Marquesas cuckoo in an early-morning concert. The mild morning air seemed green with a jungle perfume. Whatever unknown we were heading for, we felt we were returning to a lost and luxurious garden that was ours for the asking. No fences and no guards. It felt like a dream.

It was the old, overgrown royal path we followed inland. Away from the village. The red, dented ridge that soared skyward at the bottom of the valley disappeared from sight as the jungle began to close above our heads. First, young coconut palms, resembling giant ferns; then mighty jungle trees with moss-studded branches bearded by parasitic growths and hanging ropes of lianas. There were times when we could scarcely see the rays of the sun playing on the upper foliage, which was filled with hooting, fluting, fiddling, and piping creatures. There was life everywhere, although all we saw were tiny fluttering birds and butterflies and lizards and bugs pattering away from the exposed trail. We hurried to get ever deeper into this all-absorbing wilderness. We were keen on getting away as fast and as far as possible from the little cluster of village houses; where a complete lack of sanitation seemed to have brought all kinds of diseases upon the natives. We had waved to the last ones on the outskirts of the village. They waved back, shouted *kaoha*, and chattered unintelligibly. 'Good day,' we answered in their own language, 'nice, nice; good day.' Then we all laughed together. They seemed happy despite their diseases, though some of them could hardly walk, their legs thick as their bodies. The elephantiasis from which they suffered had come to the island when the white man unintentionally brought the mosquito ashore. The last we saw was a little group of women sitting waist-deep in a pool of the river, washing themselves in milky water with all their clothes on, while others filled their gourd containers with drinking water a few yards farther downstream. They knew nothing of hygiene or contagion.

Just above the village, the river water, filthy lower down, became clean and fresh. The trail followed the stream, occasionally winding across the transparent water on smooth stones and sometimes cutting into the rusty-red soil of the riverbank. At the beginning, the trail was kept wide and clear of intruding bush, but as we advanced ever deeper inland, it became narrower and we often had to use our long machete. Willy had picked our guide, Ioane, his own Marquesan brother-in-law. He had a very definite idea of the place he would show us for our future homestead: the site of the last island king.

The grandmother of Ioane's grandfather had been the last island queen before the French annexation. Through her, Ioane had inherited the part of the island we were now heading for. We had learned from Willy that, even though the people might have died out, there was not a spot on the island without an owner. Everything was divided into family property that passed on to some heir, and even if a plot of jungle was abandoned and all but inaccessible, woe to the man who pilfered a banana from another man's land. If seen, he would be reported to the village chief.

Far up the valley, where the river was reduced to a rushing stream, the trail was gradually lost. Here we left the water and the valley bottom and began climbing in a wilderness of boulders, bush, and giant trees. Hidden everywhere in the undergrowth were ruins of moss-covered boulder walls. Artificial terraces. There was evidence everywhere of a persistent fight between stubborn garden trees and the returning, omnipotent jungle. Heavy walls with boulders weighing tons had been pushed apart by the muscular roots of trees so big that the three of us together could not encircle them.

Finally Ioane stopped. A spring of cold, clear water gushed forth at our feet. Next to it was an artificial plateau so overgrown that it was impossible to get a view of the valley or an impression of the place itself. This was the royal terrace. Here the last queen had lived. It did not look too inviting, but trunks of coconut palms and of breadfruit trees could be distinguished in the chaos of foliage, and we spotted the huge leaves of bananas and as well as taro, the largest lemon tree we had ever seen, loaded like a Christmas tree with golden fruit.

If the site had been picked by the royal family, we could hardly do better elsewhere. We made it clear to Ioane that we had decided to stay, and he saluted happily and disappeared into the greenery. We had agreed, through Willy, that for a price per year equivalent to the value of an empty suitcase we were to rent the entire surrounding part of the valley, with the right to clear and build, and to eat all the fruit and nuts we could manage. In addition to this modest rent to Ioane, we were to pay a trifle in tax to the village chief.

For the first night, we had brought along a tiny tent, really an inflated bag with a zipper opening, and before the sun set, I had managed to clear enough space to put it up. As night fell upon the jungle, we used matches for the last time. From now on, we were to save our evening embers under ashes, and if the fire died out, we had to rub another into life with the aid of split hibiscus sticks. That evening, Liv baked some *fei* and her first breadfruit, as large as a baby's head and then, as happy as children, we crawled to bed in our tent below the huge, sprawling leaves. The mosquito was the only devil in our Paradise.

Sleeping in a tent is the next best thing to sleeping in the open. With nothing but a cloth wall to separate you from your surroundings, you participate in the faintest sound around, particularly so in an unfamiliar kind of forest. That night, we were to learn a lot about the natural environment that was to become our home, and yet there was much we did not understand. What made that ghastly cry as it seemed to leap across the canvas? It could croak like a toad and creak like a rusty door. Something was rummaging in a pile of stones nearby. Was it a wild pig? Higher up the valley, something hooted like an owl. And we clearly heard the mewing of a cat somewhere below our terrace.

Then we both heard someone approaching. We heard steps in the dry, fallen foliage, and the casual sound of cracking twigs. The sound came closer, stopped. A long silence. Then someone tiptoed right up on the tent. A sudden silence again, while we listened, in complete darkness, with racing hearts.

Who in the world could it be, slinking up here at night and now standing motionless outside our tent? In our minds we both saw a revengeful native with elephant legs leaning over the little tent with a fishing spear, or perhaps a heavy stone, in his raised arms. They had no reason to love the white man here, where our worst diseases played havoc, where no one could read or write, and where the cannibalism mentioned in the old encyclopedia persisted until very recent generations. If we disappeared, nobody would ever know when or how.

Lacking any weapon, I hurled myself through the zipper opening with an earsplitting yell. It is hard to judge who was most afraid, for outside was a wild white mongrel dog who was staring at the tent with a stupid expression. It became so madly frightened that it shot off like a white arrow into the bush and down the slopes, and never ventured back into our domain.

Even so, we were not left in peace. In the calm night breeze, we heard scattered gunshots from distant parts of the valley, a few of them even fairly close. This was crazy, for hardly anybody on the island but Willy Grelet could possibly possess a gun; besides, nobody could see an arm's length in the forest at this time of night. We had to crawl outside to hear better. We could see a few stars above. Then something hit the ground beside us with a tremendous impact. It resounded like a gunshot, just like the ones we had heard in the distance. Something rolled toward me: a large coconut, almost as big as a rugby ball, since it still had the thick, leathery husk outside its shell, a shockproof cover that protects the nut from cracking when it falls from the crown of the palm. As the breeze increased at night and shook the lofty palm leaves, ripe nuts loosened, and, heavy with milk, they made the silent jungle resound each time one plummeted to the ground. One of the tallest palms was waving against the stars right above our heads; one tent stay had been lashed to its trunk. In a hurry, we untied the stay and pulled the little tent away into safety. A nut from that palm would rip the tent and be as fatal as a bomb if it hit us. In the dark, I cleared some more of the bush as best I could, and we covered ourselves with the deflated tent to escape the mosquitoes as far as possible.

As the sun rose above the Tauaouoho Mountains, hardly a ray fell upon our plateau. We were well walled in by the jungle, and not even the faintest breeze could enter and brush away the mosquitoes. We decided to clear the entire terrace and build some kind of dwelling strong enough to keep out the jungle beasts.

For breakfast, I brought back an insect-eaten cluster of ripe bananas I had found by roaming the terraces within hearing distance of Liv, and when I returned, she had already toasted some nutlike fruits that had fallen from a large tree. The water in the queen's spring was cool and crystal clear, with a most pleasing taste never found in tap water, and we were as happy as the singing birds around us, and aching to get to work.

Breakfast over, I grabbed the long machete and began to swing it like a sword against the jungle. With a single blow, the sharp knife severed the juicy banana stems like onions, and even the hardwood trunk of the breadfruit tree gradually had to yield, chip by chip. Familiar fruit trees fell one by one in company with unknown jungle growth. It was a bizarre job, playing havoc in someone's garden. Liv pulled up creeping vanilla, ferns, coffee plants, and dragged down what looked like ropes with aerial potatoes attached. As the jungle roof opened above us and the undergrowth was thinned, ever more sunlight reached the ground and played on the mighty old boulder walls built up by generations of industrious hands. A soft breeze entered. When mud and rotten branches were dug away from the nearby pool with bare hands, and a short piece of arm-thick bamboo was hammered into the bank with a stone, the water of the royal spring squirted into a boulder-lined pool that made a perfect bathtub.

The trees of the lower terrace fell next, and a most pleasant view opened of the valley below, with its wealth of palms and tropical trees all fenced in by the red mountain wall on the other side, which increased in height as it ran inland from the mouth of the valley, with peaks and perpendicular cliffs. On the very edge of the red precipice, we could see half a dozen white specks moving. Wild goats. Secondarily wild, like all other mammals on these islands. Giant cane and closely packed tufts of bamboo covered some promontories at the foot of the cliffs above the palm forest. To the right, we could see a couple of miles down the valley, but the little village and the ocean were hidden by giant leaves.

As we cleared away the roots and stones, we encountered old artifacts. A variety of stone adzes and gouges, ground and polished to perfection and made from heavy, solid stone; a bell-shaped stone pounder with a narrow neck and a flaring base as perfectly curved and polished as if turned on a lathe; cowrie shells with the vaulted back sawed off to serve as vegetable scrapers; dented plaques of mother-of-pearl fashioned to grate coconut kernels; round hammer stones; and a cracked but once beautiful wooden bowl. Some, we could use. A large, comfortable stone chair had once been set out, presumably from which the king could overlook his own terrace, whereas a long bench paved with flat, smooth boulders was broken by growing palms. We repaired what was possible, and also saved a few useful or ornamental trees and shrubs, including a big bush covered with strawberry-red flowers.

After three days of hard work, most of the clearing was done, and we began cutting branches and giant leaves to build our home. We had hardly started before Ioane came up the forest trail lugging our two suitcases, which we had intentionally left behind with Willy. We did not need them. We had nothing in them that we could use. We tried to make Ioane understand that they were to remain in Willy's house. But he made it clear, with signs and chosen words, that Willy was afraid of theft. What was more, Ioane was not satisfied before he had been permitted to admire what was inside the suitcases he had lugged up through the jungle. His eyes almost popped out of his head when he saw that we who walked barefoot and wrapped ourselves in pareus were the owners of dark suits and long gowns, shoes of various colors, white shirts, pajamas, and underwear. Ties and shaving gear. Silk and a powder compact. He was so eager to carry the suitcases away again that we had to hang on to the undesired luggage and temporarily store it in the tent.

But now Ioane did not want to leave us. He made it clear that the leaf hut we were about to put up would wither away and collapse, rain would pour in, wild horses would trample on us, and all kinds of creeping things would enter. Bursting with a sudden desire to be helpful, Ioane dragged me along, higher up the hillside, into a forest of giant green bamboo. Most of the canes were as thick as a man's leg and ran up like jointed water pipes to end in clusters of long, thin leaves like fluffy ostrich feathers high above our heads. Green as they were, the thick, hollow canes could be cut with a single hard stroke of the machete. Unlike a tree, the large bamboo when cut would not topple over on one side. Sharp as a gouge, the obliquely cut end of the long and heavy tube would drop perpendicularly to the ground like a spear and cut a hand or naked foot that was not quickly pulled out of the way. With a bleeding arm wrapped in leaves, I returned to our boulder terrace with Ioane, each of us dragging a bundle of long green bamboo canes.

Liv had picked a large pile of ripe oranges, and in a few moments Ioane, although he was middle-aged, was seen atop a giant breadfruit tree; after that, he wriggled monkeylike to the crown of one of our sky-piercing palms, a feat I could never have managed even if I had not cut my arm. He left his harvest with us and made signs that he had to leave. The sun was getting low. We liked this amusing old rascal. He seemed so trustworthy, the way he trotted barefoot about in the jungle with an elegant straw hat and white shorts, apparently boasting all the time of his relationship to the last queen. Whether we understood him or not, he grinned and laughed, his brown face crumpling into foxy wrinkles.

It was barely daylight next morning when Ioane reappeared, with, on the trail behind him, his wife and four other natives. They brought us pineapples and had come to show us how to build a bamboo cabin. But before work started, everybody, including Ioane once more, had to look at the unbelievable treasures in our possession. First our tent, this amazing little roll of cloth that could be transformed in a minute into a waterproof hut. I could have kept on for hours pulling the zipper up and down, a performance invoking the utmost respect. Ioane wanted the suitcases reopened, and the excitement had no end for the spectators. To them the outside world was synonymous with the cargo of the trading schooner. A place of planks and corrugated iron. Corned beef and canned salmon. Enamel pots, matches, and underwear. Everyone shouted with joy as they struggled to peep through our tourist binoculars, which 'moved the mountains.' My little microscope transformed a blood-filled mosquito into a monster which startled Ioane's wife so much that she shrieked. And the shaving mirror, which magnified their already broad noses, made them laugh so much that they bent double as they took turns at poking their faces into it, making the most peculiar grimaces. But my wristwatch fascinated Ioane more than any other item I possessed. I had hidden it for the great celebration when our jungle home was ready and the watch could be smashed. But it was discovered before I could fulfill my dream. It was still whole when it disappeared with Ioane down the valley.

Our visitors helped us set up our new home. Our first home together. Two men with long machetes in homemade oxhide sheaths climbed the tallest coconut palms, and twelve foot long palm leaves began sailing down to their women on the ground. They split the central stems lengthwise and plaited the long fringes together like a rigidly woven mat. This would become the waterproof roof cover. Ioane and I went into a bamboo forest higher up the valley and brought back long poles of the kind my boyhood friends had used for fishing. The dark-green bamboo was first beaten flat between stones and then plaited together in an artistic pattern. A framework of wooden posts was erected, to which the ready-made bamboo walls, a bamboo floor, and the palm-leaf roof mats were fastened with strips of tough *borao* bark. The result was a picturesque green jungle home, only six feet wide and twelve feet long. Two windows and a door was cut out, and the pieces removed were tied on again with leather hinges, so door and window openings could be closed. A broad berth of round sticks was built along one side wall and covered by a thick mattress of banana leaves. Two stools of split branches, a small bamboo-covered table by the front wall window, and a tiny shelf of bamboo completed the furniture. In Robinson Crusoe fashion, we ate from huge glittering mother-of-pearl shells and drank from cups of coconut shell, with beakers and spoons of golden bamboo. Water was fetched from the queen's spring in a big calabash jar obtained by drying the rind of a gourd over fire until it became hard as plywood.

By the time Ioane and his friends left us, our suitcases were empty but for the basic effects absolutely needed should we ever return to the outside world. We also kept our plaid travel blankets, for it was surprisingly cool at night. And Teriieroo was right. There were two manmade products we could not do without: the iron pot and the long machete Teriieroo had sent with us. I had made an oxhide sheath for the knife, like those of the islanders.

Who could be richer and better off than the two of us? I would not exchange our home for any larger castle. The beautiful spinach-colored walls gradually became ever more picturesque as the fresh bamboo started to ripen and little by little golden yellow designs appeared among the green, like a chameleon changing colors, or like an artistic wallpaper coming to life. And when we opened the bamboo shutters, the entire upper section of the Omoa Valley lay before us like our own royal garden.

We had no watch, and the day was not chopped up into hours. Time took on different dimensions when marked by the sun, the birds, and our own appetites. Time never ran away from us. The days were long, perhaps because we were alert every moment in an environment completely new to us. Yet we were never bored. Each day was packed with new observations, new experiences.

In later life, when I think back on that one year on Fatu-Hiva, it feels as if it were as long as twenty of the 'normal' years that were to follow.

VI

A Taste of Paradise

THE DAY BEGAN when the spectacular, multicolored Marquesan cuckoo awoke the slumbering jungle with its resounding trumpet calls. Not losing a moment, all the other birds began their chorus of joy, one by one joining in from different parts of the valley, each with a different tune and a different love story to tell. We could not help but wake up happy, absorbing every note of this melodious morning program; which began like an overture at the opera house as the dimmed lights go up for the performance. Dawn came suddenly and yet gently so close to the equator, as if timed so as not to wake us too briskly. The growing light stole in through the bamboo-framed window opening with the last gusts of the chilly night breeze. The temperature would turn from chilly to pleasant the instant that daylight was fully turned on. The dark dragon crest across the valley was itself a spectacle as it reddened in the half-light of dawn to flame out like a cockscomb in the morning sun. For a moment, the jungle singers rose to a festive crescendo, and many of the birds seemed to be attracted to the little clearing around our cabin, where insects and worms could be readily spotted on the open ground. The parrotlike cuckoo, blue like the sky, accented with brilliant yellow and green, favored the thick foliage of the mighty breadfruit tree outside our window, while the palms seemed to teem with little fluttering singers, many resembling canaries.

No part of the day could quite match that first morning hour, when the early sun began to play along the golden bamboo plaiting, for nature was never more serene and alert, and we were part of it. Somehow, later, all living creatures became drowsy as the sun rose to pass the zenith of the sky, blessing the crowns of the tropical forest with its rays. It never felt unbearably hot, for the constant trade wind from the east carried away the mounting heat and refreshed the landscape. It was not so much the heat that made one less active nearer midday, as it was something to do with the air pressure, something that removed all unnecessary energy. The heat never bothered us in the Omoa Valley.

As the early-morning concert tailed off, we were already out of bed and down by the spring. We often surprised a beautiful wild cat with distant domestic ancestry that had the habit of sharing the spring with us. Sitting in the pool, it sometimes happened that scales of which we'd been unaware seemed to fall from our eyes, so that everything around us appeared breathtakingly beautiful. Our sense of perception seemed to be tuned in to a different and clearer wavelength, and we smelled, saw, and listened to everything as if we were children witnessing nothing but miracles. These were all everyday matters, such as a dewdrop about to fall from the tip of a green leaf. We let drops of water spill from our hands to see them sparkle like jewels against the morning sun. No precious stone polished by human jewelers could shine with more loveliness than this liquid gem in the flame of the sun. We were rich; we could bail them up by handfuls and let them trickle by the thousands through our fingers and vanish, because an infinity of them kept pouring out of the rock. The melodious dance of the stream below us, a jewel in itself, tempted us to shake pink hibiscus flowers from the branches and let them sail away, turning and leaping down the tiny rapids between the smooth boulders. They were messengers to the sea, the magician's cauldron that gave birth to all life, the perpetual purifier that cleaned the ugly village water from the river's mouth and sent it skyward and back to the hidden birthplace of our little spring.

To put our feet on the silky clay, in the soft mud, or on a hard, warm slab, felt marvelous after the refreshing chill of the spring. From the day the natives left us, our contact with nature was complete. We felt it on our bare skin. The fine climate made it a relief to strip off the clothes that the white man from cool countries was now imposing on the tropical islanders and which clung to a sunbaked body like wet paper. It felt just as good to throw off the shoes, which would be in constant need of repair from stepping from mud to sharp lava rather than from pedals to pavements. The effect was one of added freedom and added intensity. Our newly bared bodies were at first tender, like that of a reptile casting off its skin and hiding until the new covering toughened. But gradually the jungle withdrew its long claws and instead it felt as if its leaves and soft branches stroked us in friendship as we passed. Hostile to intruders, even the jungle is friendly to its own, and protective to its offspring. It was good to feel the breeze, the sun, the touch of the forest, rather than to always feel the same cloth clinging to us wherever we moved. To step from cool grass to hot sand, and to feel the soft mud squeeze up between the toes, to be licked away in the next pool, felt better than stepping continually on the inside of the same pair of socks. Rather than feeling poor and naked, we felt rich, wrapped in the whole universe. We and everything were part of one entirety.

We were living in the most luxuriant part of the valley. The abundance around us reflected man's successful attempt to domesticate the wilderness. No attempt had been made to liquidate the environment in favor of extensive, uniform plantations. The less useful trees had been replaced by more beneficial species of various kinds, which were scattered about where location and soil permitted. The planters were lost. Their domesticated jungle survived, not as the victor over an extinct enemy, but as a monument to human effort. The native people had died, not in a battle with nature, but as a result of the white man's desire to have them take part in his own civilization.

The two of us, city bred, would scarcely have survived in the jungle but for the blessings left by our island predecessors. The soil patiently continued to produce food, even though no one came to harvest it other than insects and beasts. Part of the forest behind us consisted of large stands of banana and *fei* plants. It was hard to distinguish between the two, except when they bore fruit. Both have green, sappy trunks like giant flower stems, with a cross section as big as a plate, and both are crowned by tall bundles of leathery leaves groping for the sky like palms. But a reddish tinge is apparent near the root of the *fei*, and whereas the banana has its cluster of green or yellow fruit hanging from the top of the stem like a chandelier, the *fei* has red fruit, and the cluster stands erect and points skyward like the star on a Christmas tree.

As predicted by Teriieroo, the precious *fei* or mountain plantain, which on Tahiti grew only in almost inaccessible cliffs, was growing all around our cabin on Fatu-Hiva. It became our favorite, staple diet. Inedible when raw, it was roasted on embers and eaten dipped in the creamy white sauce of grated and squeezed coconut. This coconut sauce was our only oil and served a multitude of purposes, culinary as well as cosmetic. Production was simple: we grated the nut with a serrated piece of shell and squeezed the crumbs by twisting them inside a wisp of coconut fiber. Dipped in coconut sauce, the yellow-green meat of the *fei*, sweeter than fried banana, had a special flavor of excellent quality, of which we never tired. Besides the *fei*, the forest offered us seven different kinds of real bananas, from a tiny, round variety, resembling a yellow egg with strawberry flavor, to the large horse-banana, almost as long as an arm, which had to be cooked and then tasted like baked apple.

It was unusual to come across ripe bananas hanging on the plant. When we reached for one, it was like grabbing a finger on an empty glove: it was already hollowed out by small fruit rats and consumed with the help of lizards and tiny yellow banana flies. But there was plenty for all of us. We simply collected the clusters when they were just about ready to turn yellow, and hung them unsheltered in the breadfruit tree next to our window, where they would be under our control and we could watch the sun ripen them in a day or two. Their taste was matchless, especially when compared with commercially produced bananas, which have to be picked weeks too early so as to survive the long transportation period.

The bananas hung too high above our heads for us to reach, yet I did not have to climb the slippery stem to pick the fruit. Looking like a palm with huge leathery leaves, the banana plant is merely a giant grass with a stem thick as a man's thigh and yet soft as an onion, so a single cut with my machete would fell the trunk and bring the banana cluster to the ground. The remaining stump actually did look like a truncated onion, and within a few days the inner ring rose and slowly pulled the others along upward. Two weeks later the stump looked like a green flower-pot in the midst of which stood a slender roll of leaves tall as a man. This green leaf-roll soon began to unfold like a banner, and before a year had past, a big new plant filled the old stump and a cluster of new bananas would appear, ready for another harvest.

The new banana plant crept up from the stump of the old one

with a speed that seemed both rapid and inconspicuous. A bit faster, and a passer-by would open his eyes in surprise, even in fear. A plant is not supposed to move visibly. In nature, speed of growth defines the borderline between what is natural and what is a miracle. Living month to month with the plants that fed us, I came to look upon nature as a sort of magician, whose magic wand was time. But as time began to lose its magic grip on us, the wand disappeared and we thought we could see how the trick had caught us. The same stump of a banana plant I had personally chopped off was there in the same spot every morning as I opened the bamboo shutters. But inside it there seemed to be a hidden heart, sucking and pumping. Sucking black soil up from underneath, and pushing it up into the air against gravity, reshaped as a huge plant with a heavy cluster of edible bananas on top. Soon the bananas hung there again, high above our heads.

I had never felt grateful to anyone selling me bananas in a shop, but that was almost the feeling I had toward this live plant that had labored its way up from the stump to offer me a whole cluster of free bananas, and was ready to do it again and again if I chopped it down. Nor had I thought of a banana before as a truly smart invention, a palatable meal hygenically packed and ready-made to unwrap and push between the lips.

The coconut was almost as important as the banana and the *fei* in our daily diet. The coconut palms of the Marquesas are supposed to be among the tallest in the world, and most of them were far too high for me to tackle. But there were always some coconuts that fell to the ground, and my only labor was to twist off the tough outer husk with my wooden crowbar. If the coconut was left on the ground for a few weeks, a little baby palm would put its neck out like an ostrich from an egg, while roots fumbled down in the opposite direction trying to find soil. It was not a miracle; it happened with a speed hidden from our vision, and I knew it happened because of chromosomes and genes. But it was impressive that the same soil could be made into bananas by a stump and into a lofty palm by a little coconut lying next to the stump.

Most of the food plants provided us with fruits, nuts, and root crops the year round. The orange trees, the lime, and the lemon had both ripe golden and unripe green fruit side by side, while they were still blossoming with sweetly perfumed white flowers. But the giant breadfruit trees were an exception. They marked the seasons. The trunk of this important tree was too big for me to climb unless I could reach the lower branches, and the fruit, big as a child's head, was soft and fermented by the time it fell down by itself. But I managed to secure what we could consume with bamboo rods. The islanders hoarded breadfruit in deep pits in the ground, and off season they mashed the stinking fermented pulp into the dough that had become their staple diet. One could smell it for miles, but the taste was good. We preferred to bake the fresh breadfruit over an open fire until the tough shell cracked open, and the white pulp could be broken off with our fingers, tasting like a mixture of toast and fried new potatoes.

Our kitchen was a stone-lined oven sheltered from rain by a coconut-leaf thatch held up by four posts. Here Liv would also bake *taro*, a palatable root formerly cultivated in irrigated swamps, now growing wild in the wet ground below the queen's pool. The huge heart-shaped leaves told us where the *taro* grew, and mixed in among them were also some still larger leaves of a wild plant, so big that we used them as umbrellas in the rain and as colossal fig leaves if we were surprised by islanders while bathing in the pool.

There was still more to harvest in the surrounding forest. Large, pear-shaped papayas. Small but extra-flavorsome mangoes. Wild pineapples. Tiny, red-skinned tomatoes. Pandanus, with its compound of nutlike kernels. The nobbly, blue-green *tapo-tapo*. And a single large tree with a gorgeous fruit looking and tasting like a strawberry but as large as a cauliflower.

For drinks, we had mineral water from the spring, orange

juice, lemon squash sweetened with squeezed sugarcane, and the milk of green coconuts harvested with a struggle from the lower palms higher up the hill. In Tahiti, Liv had learned from Faufau to prepare a very tasty tea from the withered leaves of orange trees. We often planned to gather and roast the red berries of a few coffee plants that grew in the thickets right behind our cabin, but got too fond of our orange tea.

It was not only the plants that had outlived their domesticators. Stealing between the trees or trampling about in the open highland were the dogs, the cats, the horses, the cattle, the sheep, and the goats that descended from former European stock, and the bushy, long-snouted Polynesian pig, originally brought in by the islanders themselves. The tiny fruit rat, a clean and happy little creature, also brought by the Polynesians in their canoes as a favorite dish, ran about in the thin branches outside our window, stealing oranges. No other warm-blooded species had reached these islands, except birds and whales. Not even bats. Snakes were unknown. But chickens had been kept on most Polynesian islands even before the arrival of Europeans. When their owners died, many had escaped into the jungles of Fatu-Hiva and survived wild. In the morning, we heard the familiar crowing of the cocks far up the valley. Ioane had presented us with some of his own domestic chickens, but we had no fence and did not dare to clip their wings for fear of the wild cat; so they flew merrily about like wild geese, slept in the highest trees, and came down only to peck what we left for food. Their eggs were laid all through the jungle, so, in spite of careful searching, we were lucky when we found one.

I spent most of the day exploring the forest for food, with baskets plaited from palm leaves hanging from a carrying stick across my shoulder. At the same time, inspired by my former professors, I was on the constant lookout for animals to save in bottles and tubes for the study of transoceanic migration and microrevolution. The idea of ever returning to the stale life of a modern community seemed very remote, yet my zoological collection could be sent back on some future schooner to be studied by someone else.

Compared with that of the continents and continental islands, the Polynesian land fauna was poor, and Fatu-Hiva was no exception. But whatever there was clearly had an important story to tell. Creeping things abounded under every stone, among the fallen leaves, and in the chaos of jungle growth. There were colorful tropical beetles and butterflies, spiders of all shapes, and an endless variety of land snails in beautiful polychrome shells. The latter, especially, would vary completely in type from one side of the valley to the other, or on either side of a tall mountain crest. Differences in animals from one locality to another were intimately associated with a corresponding variation in the plant life. Never had I seen a region where the vegetation varied so profoundly from place to place. As the months passed by, and we came to know more of the island beyond the Omoa Valley, we could see how the Tauaouoho Mountain Range was decisive for the whole pattern of plant and animal life. The distribution of peaks and passes along this ridge decided where rain was going to fall and where it was to drift. Like all of Polynesia, Fatu-Hiva lay in the trade-wind belt, and the clouds always came in the same general direction, from America toward Asia. Where low plateaus and mountain passes let the clouds blow over unaffected, there was drought; the landscape was one of dry savanna, almost desert in places, with nothing growing except yellow grass and small ferns. But in directly adjacent areas, dense jungle, sometimes really impenetrable rain forest, covered the soil, because the rain poured down every afternoon and evening. This was where lofty ridges and peaks arrested the trade-wind clouds and condensed the rising mist from the sun-warmed islands as it was forced up into cooler altitudes. Every afternoon, the central mountains were wrapped in a thick cloak of clouds, from which tropical showers splashed only over the already verdant areas.

Except when a rare tropical storm swept the entire surrounding ocean, the island rain was a strictly local phenomenon, as if the peaks were raised to milk invisible water from the blue sky, pouring it always over the same parts of the island, day after day. The flora and fauna were entirely dictated by this regular direction of wind and clouds. We never saw a cloud drifting back from Asia toward America. Asia was downhill. This was an observation I recorded on my naked skin; it was literally to blow me on the course I was to pursue into the Pacific years later.

Experience taught us that the inner corner of the Omoa Valley and adjacent mountainsides were literally impenetrable. For thousands upon thousands of years, jungle trees had tumbled down into a network of trunks and branches, lying on top of each other and overgrown with thick green moss, parasitic ferns, and flowers, and interwoven with live trees and lianas. If we tried to force our way in, we constantly fell through and sometimes disappeared completely before reaching the real floor. Climbing along on the slippery and often rotten framework, we sometimes needed a full day to traverse a jungle ravine a few hundred yards wide, fighting our way with the machete and watching our footholds in an effort not to crash through.

In these areas, even aboriginal man, with his desire to domesticate the wilderness, had achieved no success, as shown by the total absence of human traces. Nearer our home, the forest was far more pleasant and congenial, and as we began to know our way around, we moved with little difficulty. Here we stumbled upon human vestiges wherever we put our feet: mostly overgrown terrace walls and stone platforms, *paepae*, where native huts had once stood dry above the mud. But occasionally we met the former people too; their bleached or sometimes green-stained skulls and long bones could be stumbled upon in caves and crevices, and in a few areas carved slabs were set on end to mark a taboo enclosure filled with old human craniums. Sometimes, on the nicely cut slabs surrounding such burial places, were reliefs of squatting figures with their arms raised at right angles, as if meant to chase away evil spirits or undesired intruders. In a few instances, even a stone statue was erected, resembling a stout demon with giant round eyes, an enormously wide mouth, stunted legs, and hands placed on the fat belly. Petroglyphs of men and marine creatures, staring eyes, concentric circles and patterns of cup-shaped depressions were found down by the river, while high up the hill above our cabin, overlooking the valley, was a large slab sculptured into a turtle.

No archaeologist had set foot on Fatu-Hiva. One German and a team of three American ethnologists had studied the customs of the living islanders on some of the other Marquesas Islands, but not on Fatu-Hiva. And none of the islands in this group had as yet been visited by an archaeologist. There were new discoveries to be made almost everywhere I set my foot. Soon the space under our berth was packed full of polished stone adzes, ancient poi-pounders, and images made of stone, turtle shell, and human bone. Even a collection of old human craniums, which scared us one night by starting to rattle and wobble because a fruit rat had crawled into one of the neck holes.

No human voice but our own was heard in the valley. We felt like the survivors of a forgotten catastrophe. We tried to imagine the daily life of our predecessors, their work and play, their loves and their problems. They ceased to be the exotic curiosities I had visualized while I studied rows of books in Oslo or the equally tidy rows of fine Marquesan artifacts I had made notes on in the Völkerkunde Museum in Berlin. These abandoned utensils showed us that their former owners had solved their daily problems in a way natural to us, too. Their makers were not strangers at all; they were like Teriieroo and Faufau, like our friends down in the village, like us – or, rather, we were still like them, although we prefer to think differently because we have progressed in invention and changed our way of dressing. As for myself, I began to acquire a less strictly academic approach to anthropology than I had used to have at my own desk at home, when trying to digest the many conflicting theories of scholars, few of whom had seen a trade-wind cloud and still fewer who had set foot on the shores of a Polynesian island. The problem of how Polynesian tribes had found their way to these islands long before the days of Captain Cook, Columbus, or Marco Polo, began to interest me more than the itinerary of aimless coleopters and gastropods. My interest grew, moving from biology to anthropology. My task had been to study the emigration of living creatures to Polynesia. No migratory species was more intriguing than man himself.

Weeks passed. What gradually burned itself into our minds more than any artifact or animal was the feeling of being an integral part of the environment rather than something combating it. Civilized man had declared war on his own environment, and the battle was raging on all continents, gradually spreading to these distant islands. In fighting nature, man can win every battle except the last. If he should win that too, he will perish, like an embryo cutting its own umbilical cord. All other living creatures would be able to continue their existence without the presence of mankind, for they did indeed exist alone in the beginning, *without* mankind. But man could be neither created nor evolved before the rest of the global ecosystem was ready to house him. Nor could man survive in the future if that ecosystem was destroyed. Living with nature was far more convincing than any biological textbook in illustrating the fact that the life cycles of all living creatures are interdependent. As city people, we had been secondhand customers of the environment; now we were directly part of it and had the strong impression that nature was an enormous cooperative, where every associate unwittingly has the function of serving the entity. Every associate except man, the secluded rebel. Everything creeping or sprouting, everything man would spray with poison or bury in asphalt to make his city clean, is in one way or another his humble servant and benefactor. Everything is there to make the human heart tick, to help man breathe and eat.

The bamboo walls of our cabin let the jungle air pass through and fill our lungs to capacity. At home, when we had wanted a breath of fresh air, we would open a window, turn on a fan – today some would switch on their air conditioning – but who would give a moment's thought to the modest providers of the oxygen we gasp for? The baker supplies our bread and the farmer our milk, but air is just there and free for everybody. Yet the largest city would stop pulsating and industry would collapse if it were not for weeds and wilderness, for the inconspicuous roots that transform black soil into the green leaves that emit oxygen throughout the day. Smaller plants, almost invisible to the eye, float about rootless, like green dust in the surface water of all the oceans. The water around the coastal cliffs of Fatu-Hiva was green with this marine pasture, but the illiterate village people down by the river's mouth did not know the functions of plankton.

Nobody on Fatu-Hiva, and hardly anybody elsewhere in those days, except those of us trained in biology, thought of this marine dust as the foundation pillar of all life on Earth. Man himself and technical progress was the pivot of the global image then, when sea, sky, and continental forests were still unlimited.

Fresh from my biology courses, I often marveled and thought that banana plants and coconut palms, and all that can move about, from whales to elephants, owed their existence to these brainless, limbless microcreatures. Before plankton began composing the first molecules of oxygen, no fish could live in the sea.

And before the same inconspicuous plankton had sent enough surplus oxygen above the surface to form a breathable atmosphere around our planet, no flying, creeping, jumping, climbing, or walking creature could come into existence anywhere. They gave us air. Until plant plankton and their first terrestrial relatives began emitting oxygen, the winds blew only poisonous gases over land and sea. Thanks to modern science, civilized people know this. We know it whenever we come to think of it. But our knowledge has not changed our conduct. The way in which our clean valley river was polluted by the ignorant village people before it reached the sea is copied on a larger scale by modern industry and all the cities of the world. Nothing has been found too venomous or poisonous to be piped into the sea. Although Columbus has shown us differently, the ocean still seems as endless to us as to any savage, and breathing comes naturally to everybody, so who cares for marine plankton or forest weeds?

In our airy jungle cabin, before we fell asleep, we would lie and inhale the exhalation of the surrounding forest. We were mouth to mouth with the breathing greenery, the one inhaling what the other exhaled. We and everything else were united in a common pulsation, an endless machinery, a nonstop production line. The breath from flowers and plants that gives life to birds and beasts was paid back by them, and us, in daily doses of carbon dioxide and manure. On Fatu-Hiva semiwild horses, cattle, and goats paid back, generously, for the vegetarian diet the wilderness let them consume. The wild dog lifted its hind leg against the tree to make sure that not a drop of its contribution should get lost. Worms and beetles dug the ground like real farmers, preparing the soil for the blindly fumbling roots, and with the aid of dogs and cats and other scavengers, ranging in size down to the invisible bacteria, they kept the jungle clean. They cleared away rotting carcasses, and left it to flowers to fill the air with a variety of delightful, pleasure-giving perfumes.

The jungle is well-groomed, and in the wilderness beauty has the same function as the pleasant scents: to stimulate and to attract. Beauty in neverending variety peeped forth everywhere, high and low. Pleasing even the eye of sophisticated man, elegant orchids decorated moss-covered branches, and slender hibiscus trees stretched their pink, red, and blue flowers out over the stream. Tiny travelers with waving antennae and eyes popping out of their heads also apparently took in the beauty, as a simple flower would offer a minute insect attractive colors and a sweet meal for its help in carrying away a few pollen grains, thus aiding an immobile plant in its task of fertilization. Each flower had its own ingenious design that made it impossible for the insect to reach any nectar without stepping on the stamens.

We shared our pleasures with everything that moved around us, and like the rest of the ecological system we took the food that grew in the trees for a given, just like the air we were inhaling. Only at certain moments, when the jungle atmosphere seemed to wake us from a sort of habitual slumber, did we differ from our four- or six-legged forest companions by asking ourselves naïve questions. It would happen when we ate a certain fruit. We'd take a second look at the mute tree that produced it and ask ourselves how the wood of one tree could come up with a delicacy so different in shape and taste from that yielded by another. A chunk or splint from an orange tree did not look all that different from a chip from a mango or a breadfruit tree, and yet, when the same mud was filtered through them, from the roots to the branches, it came out completely different at the upper end. These simple tree trunks that fed us were in fact master cooks. The raw material they used was nothing but what a child has at its disposal when making mud pies. The best cook in the world, with access to the finest choice of condiments and spices, could not convert mud into the wide variety of supreme foods we got free from these quiet trees. If a cook could do this, he would be a magician. In the jungle, we were surrounded by magicians.

One day, as I rested with my burden of *fei* under the foliage of some flowery bushes, enjoying the play of the midday sun, I detected a speckled spider on its way down the thin silk thread that it skillfully produced for its own descent. Before reaching the ground, it seemed to change its mind and climbed quickly up to the leaf from which the cord was suspended, leaving the shiny thread to flutter in the faint breath of air. The spider seemed to be patiently waiting for something, although a single waving thread could catch nothing. I had never liked spiders. Now I was beginning to change my mind. The nymphs in this forest were dancing swarms of mosquitoes, and they bit us day and night. We had started to leave undisturbed the cobwebs spun under our ceiling; they served as flypaper. We began to bless the spiders and the friendly little lizards that hid in the thatch and helped us in our battle against mosquitoes. Where would we be if there were not some control over the population increase in the forest? Every species would multiply beyond measure, and the clock of nature would become dogged and stop. Even the spider had its purpose. It did not kill out of hate or revenge, but for its own survival as a cog in the global machinery.

The wind was almost unnoticeable, but the fluttering thread of silk was so weightless that it drifted out horizontally and was soon entangled in a twig on another bush. The spider ran across like a rope dancer. It wound up the slack with its hind legs, and fixed the end to the twig. The little creature clearly had something planned. It climbed the tightrope back to the first tree, and higher up to another branch vertically above the first. On this trip back, it had spun another long cord, which was now tightened and made fast even higher. In a while, it was back on the opposite tree again, and by a deliberate choice of attachment points, an open framework was gradually constructed, like a vertical loom, on which the busy artist could begin the intended precision weaving, all with the idea in mind that, if the job were intelligently done, it would be possible to catch a flying steak.

From different points along the upper thread of the frame, the spider let itself sink to the lower thread, spinning new threads with each descent. Landing on the lower cord, it carried each new rope end left or right to be fixed at exactly calculated points, in such a manner that all the threads crossed each other at the same center, like the hub of a wheel, as if worked out with precision instruments. I knew from my zoology studies that the silk threads so far produced and suspended were not adhesive, and the spider ran along them with the greatest freedom. But as it next took up its position at the center of the star, it was ready to put into operation another kind of gland. It began revolving, leaving behind a thread that coagulated from a viscous fluid that it seemed to try not to step on. Beginning in the center, the weaver started to walk in growing circles, releasing this glittering sticky thread until it formed a growing spiral attached to the spokes of the supporting wheel. When the spider was satisfied with the size of the web, it began to spin a tubular hideout under the leaf where the work had first started. The final touch was a cord that tightened the web and stretched like a fishing line to the hideout, enabling the owner of the trap to hold the end and feel a 'bite' as soon as something was caught in the web.

The spider was a reminder of the discussions in my childhood home. Creation or evolution? What did it matter after all? If the invention of the ingenious spider web, with its gluey strands, was the result of evolution, there must be an invisible force guiding the evolution in intelligent directions. *Creating* the evolution. The spider's product was as smart as that shrewd invention of Italian bird-catchers, who also trap flying creatures by smearing glue on sticks. But no human bird-catcher could manage to squeeze glue out of his own body. And the spider had even squeezed a silk string out of another gland that would serve for climbing, and was well aware which of the two types of thread it could walk on without being caught in its own web. Did the spider think? Did it figure out a plan for catching flies? Hardly. All spiders of the same species used the same trick to secure their daily bread. They spun the same kinds of thread and wove the same type of web. They behaved as if guided by remote control, equipped by an inherited mechanism designed to command the behavior of each species, and yet individually adjustable. Because the framework of trees and branches available to the spider and from which it could suspend its web were never the same twice.

There was a definite design behind all this, something inside or beyond the individual insect that told it how to make use of its skills. Something told the spider how to use its two hind legs for spinning rope for line dancing and climbing, and how to produce thread from the right gland, so that it did not get stuck in a mess of glue. Science has made up a word, 'instinct,' and thus we have a term for whatever it is that tells the spider how to spin a web, the wasps how to make a nest, and the beaver how to build a dam. We have invented a magic word, and with it we have camouflaged an ignorance. But whatever instinct may be, and nobody knows, instinct does not help the spider obtain its glands for the production of glue. And plants have no instinct. So we have to think up a word for whatever it is that tells the roots of a coconut to creep downward and the stem to rise up toward the sky. Gravity explains why things fall, roll, or run down. But we need a word for the other force in the coconut, the one that lifts the crown with all its burden upward. Gravity is a word we have invented to explain why the ocean clings to all the sides of a round planet instead of splashing off into space. To explain why people on one side of the planet walk 'upside down' compared to others. Iron or flesh, feathers or giant rocks, everything is dragged toward a tiny point in the very center of our planet through some truly mighty force. We have no idea what the origin of gravity is. But we eliminate our ignorance once we have a name for it. It is the same with the flowers in the forest; we feel we know them the moment we know their names. Words are useful, not least those we fabricate to hide our ignorance. To make us stop thinking ...

Primitive tribes and old civilizations had priests and sages who used the same system to satisfy the curiosity of the masses. To explain the forces that converted soil to plants. Sperm to man. Short of real understanding, they resorted to a word. Man has thought up an infinity of different words for the power evolving and maintaining the ecosystem, for the invisible forces in nature that make the rivers run, the rain fall down, the plants grow up, and man move about creating words. Throughout the ages and in all societies man has marveled at his own existence, seen his own limitations in the universe, and created a word for 'god.' And the image of god has been depicted in art with as much variety as in the names invented.

Among the founders of the first great civilizations in both the Old and the New World, the sun was seen as the creator and supreme god. The great thinkers of antiquity assumed that the sun was the father of all life. In ancient Egypt the sun god Ra was depicted as an eagle-headed man with wings and a solar disk on his head. This same concept, that of a divine bird-man with human body and a hook-beaked bird's head, was widespread in antiquity on either side of the Atlantic, from Mesopotamia to pre-Inca Peru. The gods of other cultures were often depicted as monsters, some with many arms, two heads, or an extra eye in the forehead. Since medieval times, European conquerors and missionaries have spread throughout the world the concept of the Creator as an old man with a flowing beard, dressed in a long white gown and wearing sandals. This surprised the islanders of the Marquesas group, whose pre-Christian creator-god was both male and female. They frequently depicted him in stone as small 'double-Tiki', a naked man and woman sharing one back and facing in opposite directions.

Muslims worship Allah, but make no attempt to give him a shape. And modern Christians no longer think of God as an old man walking on the clouds, but as something abstract, everpresent, and totally invisible. As invisible indeed as the forces science credits with the elaborate evolution of the species. For the boundless power of gravity. For the everpresent network of instinct. As invisible as the nameless mechanism that tells the spider how to spin and weave and never step in its own glue. Whatever it is, it is there, irrespective of name.

In our days of astronauts and molecular biology the term 'god' strikes a discordant note to many ears. It tends to bring to mind some sort of fairytale patriarch or medieval painting. It is difficult for anybody to perceive the existence of something that cannot be depicted or measured. On Fatu-Hiva, it was not so difficult. Far from any church or city, we felt no need for images or symbols. It was all there, in everything. And we were part of it whether we named it or not. Everything around us evolved and functioned, obediently and automatically, without our interference and without the interference of the outside world.

We were free to call ourselves the 'lords of this planet' and to give Latin and Greek names to the laws and forces in nature that function beyond the range of all our senses and all our instruments. While religious creeds quarrel about God's name, and scientists think up new vocabularies, it becomes ever more obvious for biologists that nature's boundless variety of successful inventions cannot be the result of mere random change, but the product of immeasurable intelligence. The more we look into microscopes and telescopes, the better we understand that we have been helped into a world too wisely evolved toward a functional ecosystem to be a product of pure coincidence. A billion molecules could be tossed into the air and never come down together in the shape of a butterfly egg or the seed of a breadfruit tree.

Hungry, I returned to the cabin, where I too had started building a trap. For some days, both Liv and I had felt as if our stomachs craved something other than fruits and nuts. Certainly the reason was the total change from the European diet. We were spoiled. Yet it began to bother us slightly. We could eat until our stomachs held no more, and breadfruit, taro, and coconuts were left on the bamboo table. We were never tired of this diet, we were just filled to capacity – and yet still hungry. One night I dreamed I was eating a juicy steak, and I was almost angry with Liv because she happened to wake me up in the middle of the meal.

Something had to be done. I took my long knife and climbed up the hill, looking for fine bamboo. It must be possible to make a trap and catch some of the prawns we had discovered in the river. They were incredibly shy, perhaps because of the wild cats, and with a flick of the tail they jumped back and were gone like a flash each time I lured them to my left hand with their favorite food, a bit of coconut meat. Even if they came close enough to stretch out their claws like fossilized mittens pinching the bait, their pivotal eyes warned them of my other hand, which was trying to catch them from behind. But, that day, I wove my first crude trap of bamboo splinters with a one-way entrance and a coconut bait. Now I too could go to my den and wait for the result.

I danced in triumph the next morning when I first found my trap squirming, full of dark river prawns, some with bodies longer than my fingers and with thick claws. I put my booty in a bag of large leaves, and brought it home to the pole kitchen, where Liv was roasting *fei*. The lavish jungle feast that followed was forever to remain a milestone in our culinary experiences. We first devoured all the peeled tails and bodies. Then we cracked the juicy claws with our teeth and flushed the contents down with lemon juice. Now the *fei* in coconut sauce tasted better than ever, and we really felt happily satisfied. A new source of food was available. There was an endless quantity of prawns in the river, and even some very tiny blue fish wriggled into the traps along with the delicious crustaceans.

The sun shone as before on the golden walls of the bamboo cabin. We missed nothing. We certainly did not miss civilization.

VII

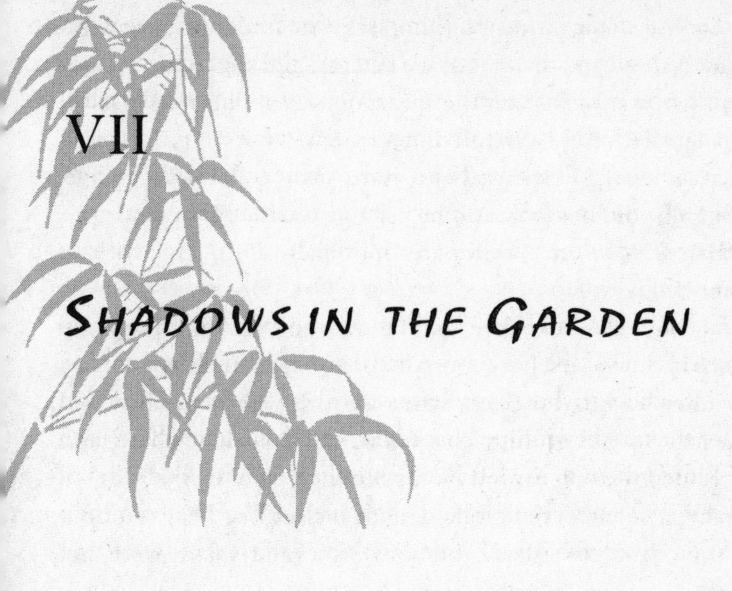

Shadows in the Garden

THE MIDDAY SUN GLITTERED in the palm leaves high above my head. Nature rested when the sun was high and the shadows short. Total peace. White doves symbolize peace. Now they were circling alone and silently around the tall crown of the coconut palm that shaded my bent back as I fumbled in the stream to fix my bamboo weir between slippery boulders.

Suddenly I realized I was not alone. I heard a noise and crept ashore to wrap a loincloth around my waist. The bushy black head of an islander with a slender spear emerged from among the ferns, bent double and trying to make no noise. He looked more negroid than Polynesian, with a broad nose and curly hair. His skin seemed almost black in the sun.

I saw him first. He was busy looking down as he waded slowly along the stream, chewing mouthfuls of coconut which he spat in the running water in front of him. Then he lifted his spear with lightning speed and drove it down between the boulders. A good-size prawn was spiked on the spearpoint until put in a calabash container that surely was full of many more.

'*Kaoha nui*,' I shouted to him in the local language, although he did not look as if he belonged to the Marquesas.

'*Bonjour, Monsieur*, ' he replied in French, saluting me with an unexpected bow.

His name was Pakeekee, and he was the Protestant pastor on the island. I had never seen him before, as he had been over in a side valley trading for copra when we came. Teriieroo had sent me a note to give to him, but Ioane had shaken his head and caused me to believe he was no longer on the island.

Incidents that were to follow made it clear that he was indeed very much on the island, but Pakeekee and Ioane were not friends.

Pakeekee had been sent from Tahiti to preach the Gospel to the islanders on Fatu-Hiva. On Tahiti almost everybody was Protestant. As in northern Europe. In the Marquesas, the Catholics were totally dominant. As in France. We had never thought there was much difference, and were amazed at the delight of Teriieroo and his neighbors in the Papeno valley when they found out that nearly all Norwegians were Protestants by birth. When Sunday came, Teriieroo's wife had put a huge plaited hat adorned with heavy shells on Liv's blonde head, and Teriieroo made me tighten a tie around my neck. Thus dressed, we had marched with everybody else in the valley into the church. A Protestant church.

On Fatu-Hiva, nobody had asked us about our religion, and nobody seemed to care. Until I met Pakeekee. We were immediately caught in his net. Pakeekee asked us to dinner, and requested two days to prepare the meal. His party lasted for days on end. Never had we eaten like that. We had hardly staggered away from the table before we were fetched back, morning, midday, and night. Our host and his family had made ample inroads into the chicken coop and the pigsty, up the trees and into the ocean. And as we entered his one-room plankhouse in the village, smoke and a delicious fragrance from the open kitchen shed whetted our appetites and delighted our eyes.

The table was set for four. A tall fellow with a friendly, funny face was seated with the host and us. The hostess and the children sat on the floor. It seemed to us as if the entire village population were climbing on top of each other to get a glimpse through the open window. And for a moment we thought we recognized Ioane, who seemed to have a long face.

'Are there many Protestants on the island?' I asked, to be polite.

'Not so many,' was the answer. 'When Father Victorin visits Fatu-Hiva he gives everybody sugar and rice.'

'But how many are Protestants?' I repeated.

Pakeekee began counting his fingers: 'One is dead,' he said. 'And then it is me and the sexton.' He nodded toward our eating companion at the table, Tioti, and hurried to add: 'There was one more, but he moved to Tahiti.'

By the third and last day of the party we had finished the food. It was Sunday, and as we struggled to our feet Tioti went to fetch a huge conch trumpet. He went out on the one and only village road and blew three long trumpet blasts that echoed between the hills. Then he came back, and we all sat waiting. This ceremony was repeated twice more. After the third trumpet blast, the pastor and the sexton rose and told us they had to go to church. Next to Pakeekee's plank cabin was a tiny hut of plaited bamboo, like ours but without windows. We were never to see the inside. That, we learned, was the Protestant church. Our two table companions disappeared into it alone. We were not invited, and started our long walk along the trail up the valley. Looking back, we saw the big white plank-house with a spire and corrugated iron roof closer to the boulder beach, which we had noticed when we landed. That was the Catholic church. But it seemed, surprisingly, abandoned.

Days and weeks passed. One late afternoon, Liv and I were relaxing in a big pool in the stream after a tough day climbing the hills in search of fruit in the upper valley. We sat chest deep in the water, cascades gushing over our shoulders. And we treated our palates with juicy mountain mangoes. Large umbrella leaves spread out over our heads and shaded us from the strong sun. Life was almost too pleasant to be true. Birds and butterflies. Hibiscus flowers alongside the stream. We asked each other jokingly if we were missing something, if there was anything we would care to buy. We rubbed jungle resin off our hands with pumice and washed our weary feet with coconut oil.

The joy of mere existence still filled our limbs as we grabbed our baskets and headed up the short trail to our home.

A lone rider came up the valley. Tioti. He looked pale and drawn. He brought a piece of pork as a gift from Pakeekee and asked if the pest had not yet reached us. For the pest had reached the island with the schooner *Moana* that had just called in for copra. Many had died. Tioti spoke with the voice of a sick man, turned his horse and rode down the valley again.

For a moment we were tongue-tied. We looked at each other, at our hands, at the green parcel with the meat. Gone was the feeling of Paradise. The sickness had reached the island. It merely needed more time to spread to our lonely hiding place.

We washed and scrubbed again. We burnt the banana-leaf wrappings and roasted the meat on the fire. We ate without enjoyment, sat quiet for a while and looked at the moon. Then we crawled into bed on our bumpy log berth.

*

More days passed. We still waited for the pest. We had no medicine. None whatsoever. But we did not have to wait long, and when we felt it coming we expected the worst. But as the days passed I felt nothing more than a sore throat, whereas Liv at short intervals ran for the forest. That was all. What a pest! Only a mild influenza.

But we were soon to get another impression. We had resistance. The islanders had none. One day Tioti came staggering by foot up the valley and knocked at our bamboo door, hat in hand. His skin seemed yellow, he coughed, and there was fever in his eyes. We could hardly recognize the jolly sexton.

He asked if I could come down to the village to photograph his last son. Now they were all dead. From the pest.

In the village we met real tragedy. Piglets were being slaughtered and carried about. There were funeral parties in most of the homes. Nobody was seen in colorful pareus. All seemed black. Even the village was dark and sombre. In Tioti's modest hut, a little boy neatly wrapped in white lay dead on a green pandanus mat in the middle of the floor. Tioti's other children – we never learned how many – were already buried. We were to photograph this remaining child, as Tioti wanted a picture like one he had seen on the wall in Willy's house.

There was coughing and lamentation in the houses. Instead of opening the shutters for fresh air, the villagers closed all openings to keep the pest out. We tried to preach hygiene, but it was easier for them to believe in an evil spirit, invisible because lacking flesh and bone, than to believe in a virus, invisible because of its size. If so tiny, how could it kill big men? They had been converted, and believed in angels and devils the size of human beings, but neither Protestants nor Catholics had been made to believe in invisible germs.

Airy huts in an old fashioned style like ours in the upper valley were hardly to be seen any longer down in the village. Bamboo and thatch could be picked for nothing and did not stimulate modern trade. Compact walls of imported planks and corrugated iron on the roof testified to cultural progress, and so did the new habit of depending on imported rice and flour for food. And, at least on Sundays and when the pest arrived, cultural 'progress' required one to button up from neck to ankle in imported cloth, leaving only the feet bare. Returning home on bare feet, they crawled straight to bed, young and old packed together on floor mats, the sick and dying coughing all over the rest.

We knew before we came to the Marquesas that an estimated 90 percent of the once-healthy islanders in this group had been wiped out by the new health problems brought by white man. Venereal diseases, tuberculosis, smallpox, leprosy, and elephantiasis had literally decimated the population. Now a common flu, on top of all the rest, was enough to cause disaster.

We did not feel proud of the outside world we represented. We have done our best to make these islanders dependant on us, or rather dependant on our products. They would make no contribution to the world's progress if they continued dressing in straw skirts and eating what came from the trees. They must work for us and bring us copra in order to be able to buy the blessings we offer them. And if they help us earn enough, there may one day be funds to give them medical care and lessons in hygiene.

We left the village, heading for our wonderful little cabin with a feeling of guilt. We, as Europeans, should have been forced to sleep in the hot and stuffy houses white man has made the natives build. And they, as Polynesians, should have maintained the privilege of sleeping in our kind of airy hut. How can our own kin be proud of a white skin when our shadows fall so blackly on others?

We were filled with dire impressions as we crawled to bed on top of the rustling dry banana leaves in our moonlit hut, once more alone with the jungle. Here in the wilderness we felt safe, far from all danger of infection.

We thought of Ioane. We had seen him at work as the village carpenter. When someone got too ill to sit upright on the floor mat, Ioane came and began patching together a coffin right in front of him. The islanders did not fear death. They had always believed in an afterlife. We often found bits of wooden bowls or other grave goods in ancient tombs. We never saw a live person on Fatu-Hiva wearing shoes. Their feet were too big. Even Pastor Pakeekee was barefoot during his three-day party albeit with black dinner jacket and shorts. But new shoes were forced on to the feet of the dead. Shoes that no longer pinched. We also witnessed one man being buried with his accordion and another with a deck of cards.

That night, with a vision of all we had seen down by the sea, we began to feel a sneaking uneasiness. Things had not quite worked out as we had anticipated. We had seen an entire village suffer for a lack of links to the modern world. Lack of medicine. Lack of knowledge about bacteria and hygiene.

'Medicine is civilization,' said Liv laconically. 'Without a microscope, Armauer Hansen would never have found the leprosy bacteria.'

No use denying that she was right. Of course medicine is one of the blessings of civilization. Nature had bred the bacteria. Man had invented medicine and pesticides in self-defense.

Many of the ailments and health problems of modern communities were brought upon man by himself, due to unhealthy food and lack of exercise. We agreed on that. But the unhealthy life style of one individual did not create a virus that would attack others. Contagious pests were the products of nature. We agreed on that also. But if nature had brought forth mankind, why allow contagious germs to reduce our numbers?

Pests are there to create balance in nature, I explained to Liv, and told her about the topic that had fascinated me most during my biology training. Life on Earth was made possible by an ingenious mechanism not yet fully understood. Nature had certain unwritten laws, and a host of potential police forces to reinforce them if any of the members of the biological community caused damage to the welfare of the whole. One truly basic law was the law of equilibrium. No species is allowed to multiply to such an extent that the survival of the others is threatened. Each individual species that has come into existence on Earth is designed as an indispensable cogwheel in the global clock. When I was a biology student, I felt that by dissecting the animals to study their parts under a microscope we behaved like children taking a watch to pieces to find out how it worked. We were excited and proud when we discovered how things functioned: how the bones in an animal's ear were linked to transfer the vibrations of the eardrum to the nerves that fed the sound into the cells of the brain; how grass passed up and down the throat and between the five stomachs of a cow chewing the cud; how the living cells were composed, and how they split to multiply and form complicated organs. But like the boy who opened his father's watch, we humans marvel at what we discover despite eliminating what we do not like before we find out why it is true.

As a zoology student, I had constantly been reminded of the similarities between the organs of man and beast. Man seemed to have inherited all his organs from the animal kingdom. To my professors, man was just another species of animal. What mainly separated us from the other mobile species were our mental abilities. When mankind came into existence in the wilderness, naked and unarmed, we survived, winning the battle for the fittest solely because of the superior construction of our brain. The human brain is the most ingenious and supercomplicated invention ever put together from molecules. We are still short of a complete understanding of the interplay between the myriad pyramid cells and innumerable other minute parts that work together to help us think, talk, see, hear, feel, smell, taste, remember, have moods and feelings, and make up our minds about what to do next.

We got our brains without blueprints, and nobody told us how to use them. There was nobody around but the beasts. The smartest among them were the chimpanzees, whose greatest sign of individual intellect was their ability to grab a stick and poke with it. It took more than that to design a human brain.

Whatever it was that triggered the evolution, it required something greater than a stroke of good luck to guide it in the right direction so that, in effect, we would end up with a human brain inside an animal cranium. Whatever it was that guided this impressive alteration in man's favor, it took superhuman skill and ability to provide us with our brains. The creative force behind the atoms that built human genes did not show up as any part of the human brain. When nerves grew conductors that brought impulses from all our organs to our brain, the complex wiring system this required limited our ability to perceive anything beyond the channels and the wavelengths that matched the contact points at the receiving end inside our cranial casing. The brain is composed of molecules, and designed to register waves from other combinations of atoms. If atoms were produced by evolution, the forces of evolution antedate the atom. Something antedating the atom cannot be composed of atoms, and hence, no matter how important, it cannot be registered by any brain.

Man's brain grew like an embryo within the magic womb of nature. Instinct told us how to use it. Instinct was the magic force present in all mobile life. Instinct was active everywhere, and over time showed all born creatures how to use the ingenious organs they were blessed with. Instinct was the driving force that helped every organism that grew in the wilderness to eat and breed and multiply. Without instinct, animals would stumble over their own legs, eat the wrong food, and never understand what was in front and what was behind.

If instinct also was the invisible force behind evolution, then Instinct should be spelled with a capital *I*. In this way, we would have trapped in the world of words the evasive, invisible, and immeasurable but ever-present and omnipotent power behind the ecosystem. We would have the first scientific term for the creative force that triggered the evolution of the species and guided it through evermore complicated inventions, its ultimate goal being the creative human brain. Animals were forever to be guided by instinct. But mankind was given also the freedom to use their brains as they pleased. And we have used them from the time we received them to clear a space in the wilderness for ourselves. And, for an ever-increasing number of our own kind, with little or no regard for others.

As a student, I had witnessed how the law of equilibrium functioned in the animal world. In certain years, the field mice and mountain lemmings multiplied so much that they threatened the vegetation and polluted the streams. In the following year, the fox would suddenly breed abnormal litters, and the hawk hatch above average numbers of eggs, until the excess number of little rodents was devoured. Then the excess number of foxes and hawks would begin to disappear for the lack of food.

Man is an important cogwheel, but not the only one. He cannot rotate alone. If we outsmart the mechanism of equilibrium by assaulting our environment with arms and defending ourselves with all manner of medicines, we release a trigger against ourselves. When man moved about in family groups, far apart in the endless wilderness, the cholera bacillus had no chance to start epidemics. But when mankind crowded together in cities, and his refuse began piling up outside the walls faster than nature could convert it, then bacteria and parasites got their chance. And where lions and tigers had failed, the hordes of invisible micro-police managed to cut back the numbers of human beings that would otherwise, first, have overfilled the world and, next, have perished from starvation.

'If not from war,' Liv inserted. As a student of social anthropology she had been told that man's aggression against fellow humans was an instinct grafted into the human species by nature. Monkeys never formed troops to wage war upon their own kind. But man invents the most inhuman armaments to assault others so like himself that uniforms are needed to distinguish between friend and foe.

Perhaps with weapons, sprays, and medicine mankind could one day stop all the other species from fulfilling their duties of sustaining a balance in nature. Then man himself would be nature's last resort for help in reducing the overproduction of mankind. The alternatives would be war or enforced birth control.

I was completely sure that a second world war was brewing. My mother had tried to convince me, with her Darwinian faith, that man was getting smarter with each generation. There would never be another war after the big world war that ended in 1918. Science would save the planet from overpopulation by convincing all mankind to practice birth control. My mother was the first person I heard passionately preaching birth control.

War, pests, or birth control? Liv and I lay thinking and talking until the multitude of winged creatures outside began their morning chorus, announcing the arrival of another day. Out there nobody thought of birth control, war, or pestilence. There was something wrong somewhere in the way life on this marvellous planet was heading.

Some days passed before our part of the valley had another visitor from the village. Tioti came with fresh fish from the ocean. Tragedies are soon forgotten in Polynesia. Life in the village had returned to normal, and people had been out fishing. Liv prepared an exquisite meal of fried fish and taro, while I invited Tioti to sit down with me on the ancient stone seat with the backrest that the former king had left on our house platform. Tioti began scratching nervously. This was always a sure sign that something was wrong. And sure enough: the pest that came with the schooner was gone, but the schooner had set ashore Père Victorin. And he and Pakeekee were not friends. None of the

Catholics spoke to the two Protestants as long as their priest was on the island. Now Tioti had come to help us hide until Père Victorin had left. We could move up into the highland, or he could take us to the next little valley in his outrigger canoe.

I laughed and told Tioti that in my country Catholics and Protestants were friends, there was very little difference. Both prayed to the same god in the name of Christ, Son of the Virgin Mary. Tioti knew, but went on scratching his back. We had to hide. Not because of Père Victorin, but for the other islanders who thought they would do him a favor if they could drive us from the island. Père Victorin believed we were missionaries. Why else had we come to this island?

We had heard of this little priest who had spent all his life traveling between the islands in the Marquesas. Pakeekee and Tioti tried to lure away members of his congregation each time they blew their conch trumpet. We could well imagine the unhappy feelings of the priest, who did not know why we had come and why we were friends of Pakeekee.

We recalled what we had heard from the captain of our schooner when we called at the Takaroa atoll on our way up. Two Mormon brethren had settled there and already succeeded in converting the entire population to their faith. Except two. The two were the Protestant pastor and the Catholic priest. Each of them remained alone in his own big plank-church.

We asked Tioti to tell the priest that we had come to collect insects, not souls. But Tioti gave us a last warning before he left: Old Haii, with elephantiasis in his scrotum and both legs, had once urinated in a calabash of orange beer and offered it to Pakeekee as a punishment for his heresy.

Tioti managed to confuse us. What to believe? It would be foolish to ignore his warning. But what to do? One thing: find out the truth for ourselves.

Next day we walked down the trail toward the village. We had good friends in Veo and his beautiful wife, Tahiapitiani, who lived in the first hut before the village proper began.

But they were indeed not happy to see us, visibly disturbed and worried that someone might find us there. Finally they made it very clear that Tioti had not exaggerated. And as proof of a clandestine friendship they confided that a man in the village kept a large female scorpion in a box, with all its poisonous youngsters. It was to be let into our cabin.

Never before had there been scorpions in the Marquesas Islands. But a pregnant one must recently have come ashore with the cargo of some schooner, for I had collected a few large specimens among the beach boulders.

I exploded in fury and rose to rush straight out and talk to Père Victorin. But Liv calmed me down. We had to keep our heads cool. After all, there was no way of escaping from the island.

All night long we lay awake listening for unusual sounds. We got up long before dawn, pulled the old tent from under the bed and rolled it up in a sack, together with our two blankets and the iron pot. We had decided to follow Tioti's advice and take refuge in the uninhabited high plateau of the inland.

The dogs had awakened the entire village before we reached Tioti's hut at dusk, where Liv was given Tioti's mare and I Pakeekee's stallion. And with Tioti as guide we stole away down to the beach and up a steep trail to the inland mountains before anybody except the dogs knew what was happening.

The air got cooler as we climbed, and as the view widened over the sea and the dark valley, the sun rose with us. Once again we felt overcome by a sense of boundless freedom and happiness. The problems were left behind down in the deep valley. For some time we had felt as if choked by a deluge, perhaps affected by the heavy jungle heat. Now we again believed that we shared in the ownership of a whole world. No walls, no boundaries said: *this is mine* and *that is yours*.

Soon we were so high up that it felt a bit chilly; there were no more palms, no fruit trees, only ferns and grass on ridges and hillocks between ragged peaks and jungle valleys. The ocean gradually became endless, visible in all directions, curving around our island, all around our planet. We no longer felt like Adam and Eve, but rather like Noah when he landed on Mount Ararat. We were saved from the flood, here on top of a mountain with water around us everywhere.

It was unbelievably beautiful in the open landscape between the rocky red peaks, where mountain goats were climbing. But there was no food to satisfy our hunger, only a few scattered shrubs bearing unripe guava fruit. No one had ever lived up here. Still, this was a safe refuge for descendants of formerly domesticated animals that had run away and become wild as their masters died. We scared up little flocks of wild horses, their tails long enough to touch the grass.

But the cool air had given us an appetite, and the food we brought from the valley was soon devoured. Hunger drove us down from the mountains after a few days. We were longing for fish, meat, coconuts, and breadfruit. And for our own river prawns, our own *fei*. We thought of the little wild pigling we had caught among the ferns but left to run back to her angry mother. Forget the box with scorpions. Food was all I thought about as we rode in silence along the trail that led like a narrow red carpet down the hill.

But we felt renewed. Hungry, but also full of appetite for a new attempt at defying this newest problem. Down in the valley, I surprised Liv by setting out directly for the house of Père Victorin.

It was hard to say who was most surprised. I had not expected to find a timid little man with elephantiasis in both legs. There was nothing evil about him as he rose and came toward me in a black gown that seemed far too big. If his eyes were sad and his smile weak, this reflected a lifetime of hardship and self-sacrifice dedicated to the sick and the dead – with one single interest: to save any soul from being lost that he had already converted. He had every living person on Fatu-Hiva listed as a member of his congregation.

After some introductory remarks that reflected mutual uncertainty, we had a lot to talk about. He had never heard of people collecting insects, or who of their own free will came to live in the wilderness. Père Victorin had traveled alone in the Marquesas, going from island to island, between the sick and the dead, for thirty-three years. Always alone. None of the islanders had ever offered him personal friendship. He had slept in their dim huts and eaten their ill-smelling *poi-poi* and other dubious food. He felt nothing for the palms and the flowers. For him the island nature was oppressive. His Fatu-Hiva was the white-painted church and his own little plank cabin behind it. His pride and passion was a black-covered book in which he listed every soul on the island he had saved from the eternal darkness. Only Pakeekee and Tioti were missing.

We left the little man with the black gown and the elephant legs with a mixed feeling of relief and pity, almost reverence. In a sense we parted as friends. He saluted us with a faint smile, standing motionless in his own doorway watching us taking the horses back to his rivals, Tioti and Pakeekee. We wished we could have left the horses right there and walked straight to our own home in the valley.

VIII

Jungle Rain and Ocean Waves

THE RAINS CAME. Not as a short afternoon shower, cooling the air. Not a cloudburst, threatening to send a deluge down the valley. But sneaking in as a thief stealing the sun and hiding it in a big wet sack. In the absence of the sun, the thick sack of clouds hung over us day and night and sprinkled a light rain down upon palms, ferns, everything. And it was dripping everywhere, although not a drop leaked through the palm-leaf thatch. But warm tropical dampness sieved through the plaited bamboo walls. Our blankets got heavy with humidity. We could no longer hear the faint murmur of the river in the valley. Day and night our ears were filled with sounds of water dripping, trickling, purling, gushing, and splashing from all directions. And the air smelled of mold and fungi rather than of flowers and spices. Our cabin became a ship in a sea of jungle mud.

There was little but our empty stomachs to tempt us outside the shelter. Only for some short moments did the sun peep forth. It seemed hotter, as if its rays were turned on to maximum to dry up the jungle mud and misery. We comforted ourselves with one single advantage: we could see if somebody had come to our cabin at night or when we were out. Nobody could sneak up and put scorpions into our bed without leaving visible traces in the mud.

One week passed, and two, and three. The rain fell and the mud rose. We stopped all longer excursions and kept to the familiar area close to the cabin, where we could better attend to some plagues on our feet that had begun to bother us. Liv got big boils on her legs. When I scratched or cut my feet and ankles, I got open sores that would not heal.

One day Tioti came up to see us. We had not noticed before that one of his legs had begun to swell. Elephantiasis. This horrible disease was spread by the mosquitoes that inject a filaria when they sting. Tioti could assure us that we had nothing like his ailment, but *fe-fe*, a common problem that would pass if we used the right herbs. He taught us to boil yellow hibiscus flowers and cover the wounds with the hot paste. This we did. But Liv's boils merely burst and turned into open sores like mine. And the sores increased in size. Some days we remained in bed. On our backs. For if we turned on our sides the cabin smelled unpleasantly of mold from the bamboo walls, and if we turned on our bellies, it smelled even more strongly of the mold in the leaf mattress.

All sorts of insects began invading our home to escape the mud outside. Tiny yellow ants came marching in armies up through the floor. They found our food, even if we hung it in coconut bowls suspended like laundry from a thin string of hibiscus hemp stretched across the room. Once Liv turned over the banana leaves in our bed, she discovered three mutually unrelated ant societies, their disturbed inhabitants running in all directions to save their eggs.

It did no good to keep the poles under the bamboo floor in pools of water, for the insects flew in or fell down from the trees. All this water created a paradise for the mosquitoes, and millions of larvae wriggled in the jungle pools. And as soon as nature equipped them with wings they seemed to take off on a straight course for the bamboo cabin. Inside or outside, we were always surrounded by a thick cloud of hungry mosquitoes.

One night, when Liv had been rolling back and forth on the floor, in desperate pain from mosquito bites, we gave up. We could not win. We could not survive further if we were to be literally tortured as blood donors for these hordes of winged devils.

Limbs and bodies burning as if dipped in formic acid, we limped down the path to Willy's home by the sea. Cash payment secured us a major portion of his own mosquito net, and to keep the mud off our infected feet we bought two pairs of the large tennis shoes the village people bought, to be worn only by the dead in their coffins. We hurried home and slept until afternoon the next day.

I was wakened by a terrific scream. When Liv finally crawled forth from the mosquito net, she had tried to put on a shoe, and out came the largest spider I had ever seen, hanging on to her toe. It looked like a mouse with eight hairy legs as it dropped, ran across the floor, and escaped. Liv's toe had marks from the spider bites. I hurried to squeeze out blood and rubbed in lime juice. We expected the worst, but Liv assured me that the worst had been the moment she felt the bite and saw what it was that hung on her toe.

Something we never were to see attacked our cabin and gradually became the worst of our tormentors. This invisible enemy horde did not care if window shutters and door were closed; they entered through the door, the shutters, the very walls like ghosts. And once inside the bamboo, there was no way to get them out. We noticed nothing until a very fine white bamboo dust began to fall from innumerable tiny insect holes in the walls.

It fell like snow on the floor, like flour in our bed. It felt like powdered sand in the nostrils and between the teeth. Ioane had never told us that a house built from green bamboo would straightway be assaulted by jungle bugs. It is spectacularly beautiful, but will never last long. We should have built our home from ripe, hard, yellow bamboo. Too late. Soon we would be able to push our hands through the walls.

On one of his rare visits Tioti brought us a big chunk of freshly caught swordfish, and news of new problems in the village. Willy's store had run out of rice and flour. Unless some schooner called soon with a new supply, people would starve. They were now accustomed to eating imported grains. They depended on the schooner as much as the schooner depended on them. Coconuts were not bread to them any more, but money. It filled sacks, waiting for a schooner. Stinking sacks of sun-dried copra were piling up outside Willy's house, filling the damp breeze with a nauseatingly sweet smell, penetrating far into the valley.

No schooner came. We started to hear voices around us in the valley. People came searching for food. Even Ioane came, claiming he had the right to harvest any fruit or nuts on the land he had leased us. Soon we found even the tallest palms around our cabin empty of coconuts, so none could be expected to fall to the ground in our neighborhood. But I knew the upper valley better than most of those who came inland from the bay, and we still found coconuts and tried as before to catch river prawns. Yet even the prawns tended to disappear in the flooded river, and nothing seemed to remain but the mosquitoes. The jungle mosquitoes gradually became so terrible that the visitors from the village were literally driven back to the open land by the beach, where the wind swept most of the flying hordes away.

Time came when the two of us in the lonely cabin had no other thought but to escape from the island. Escape *any*where. To any place where we could breathe without swallowing bamboo dust and mosquitoes. Where we could fill our empty stomachs with the many things we hardly dared to think of.

Hunger drove us down for a visit to Pakeekee. There we were to learn that the entire village was desperately waiting for a schooner, but none arrived. A woman had stepped on a sharp fishbone and a tiny wound had now got infected and spread in all directions. She had *fe-fe* like us, but the disease had spread out of control. Her foot was flesh without skin. Everybody realized that her leg would have to be amputated, if only she could be taken to Tahiti.

We returned to our cabin and waited for a ship until we fell asleep. We continued to wait from the time we opened our eyes. Then, late one evening, as we sat chewing coconuts on either side of the bamboo table, we heard the distant blast. A deep sound as from a big ocean steamer. It was too dark to find the path down the valley. Obviously a steamship! We hardly closed an eye that night. And long before dawn we pulled from under the berth the suitcase with the civilized clothing we kept in store for a possible departure. It was a strange feeling to get fully dressed. Awkward. The clothing smelled moldy, but it brought to mind many pleasant memories of a different life.

Liv giggled, and I burst into laughter when she found her little travel mirror so I could see how to tie something as stupid as a tie around my own neck. It disappeared completely behind a wild, chestnut-colored beard. My shaving kit had long since disappeared down the valley with Ioane. I stared at the sunburned face of some crazy Viking with long blond hair flowing over a white-collared neck. Liv presented a better picture, especially after she brushed her masses of golden hair down over an elegant dress. It was comical to see our own tanned faces in this civilized attire. We bent over and the jungle resounded with laughter.

Overcast but no rain. We pulled sacks, which had been

intended for our collections, up above our knees and tied them on as protection against the mud. Then we staggered off into the valley. Before we reached Veo's house we pulled off the sacks and paraded down the village street in all our white man's elegance. Our getups made a tremendous impression. Once again we felt on top of the world as we headed straight for the bay to visit the ship.

But down on the beach there was nothing to see. Nothing but the bluest ocean stretching to the horizon, with the usual pattern of whitecaps. The bay was empty. No sign of any ship. Not even a tiny schooner. The village population had seen the lights last night. A steamship had passed – far out in the ocean.

To make things worse, Tioti whispered into my ears that Haii with the elephant legs was on his way up the valley with his collection of scorpions. We hurried back through the valley. As we reached the last huts and looked for the hidden sacks to pull over our legs, we discovered Haii in front of Veo's house. With rags hanging around his waist, he rose behind a huge wooden bowl from which we had once eaten *poi-poi*. In one hand he held an ax and in the other a long machete, and with the look of a madman he waved these arms in the air but made no attempt to pursue us. There were no footprints but our own further up the valley.

In the village, the situation became unbearable. Old and young were so accustomed to an imported diet that they got stomach trouble if they ate only the former island diet. Hardly anybody had maintained the ancestral custom of storing reserves of fermented breadfruit in the underground *poi-poi* pits, although they all loved *poi-poi*, which could have fed them until rice and flour arrived. They literally began to starve for lack of rice and flour, as they claimed to get ill if they only ate pork and fish. No schooner. No rice, no sugar, no flour or cans of biscuits in Willy's tiny basement store. The last schooner had not brought adequate supplies, anyhow. Willy had paddled to

Hanavave valley by canoe, hoping to borrow supplies from a Chinese who had a store there. But the man had closed his store and left for the highlands to go goat-hunting.

We too, paddled to Hanavave in this period. In fact, in Tioti's tiny dugout canoe we, with him, paddled all the way up the sheltered west side of the island almost to the north cape, where the unimpeded ocean swells from South America rose in wild cascades against the rocks and frightened us from rounding the cape.

Tioti had made the canoe himself by scooping out a log and providing it, in the Polynesian style, with an outrigger to obtain stability. The vessel was twice as long but only half as wide as a bathtub, barely permitting us to squeeze in with some effort.

To get maximum shelter from the trade wind, we kept close to the rocky coast that rose vertically above us and sank as steeply into the clear, glass-green water. We were fascinated by the incredible beauty of the tropical valley that opened at intervals, only to disappear again as the rock walls closed. And as the slow swells rose and sank, the seaweed waved up and down on the submerged stone wall beside us, rhythmically exposing the most colorful rock garden of sea anemones, in which marine creatures of all colors and shapes crawled or swam about.

But among all this extravagant display of beauty and wild imagination, an ugly little pitch-black fish the size of a finger stole the show. Its head was big but its tail strong, and each time a swell rose along the cliffs, thousands upon thousands of them hurried to jump up as high on the dry rock wall as they could. There they would hang on by a suction disk, jumping about like legless frogs trying to get out of the water.

Had evolution on our planet stopped? Millions of years ago fish had crept out of the ocean to evolve into amphibians and reptiles. Why did these intrepid fish have so little success? They had certainly struggled for eons to jump as high as possible above water level, but in the end fell back into the element to which they belonged. They all remained alike, cast from the same mold. None of them had been able to develop lungs or characteristics other than those allotted to their own kind.

One night we rode in on the surf to sleep under open sky on a boulder beach in front of a cave. Inside was a subterranean lake known as Vai-po. 'Water of the Night.' The place was cursed by taboo, and legend had it that by diving inside one could find a tunnel to a dry cave, where the skeleton of a medicine man was sitting at a stone table. Dense underbrush and fallen rocks forced us to sleep on the water-worn dry boulders beyond the reach of the surf.

I was awakened by the ghastly feeling that the cold finger-bones of a skeleton were touching me all over – my face, my waist, my feet. I even heard the rattling of bones. I sat up, hitting and fighting at the air, striking nothing and seeing nothing, as it was overcast and pitch black. The others began to whimper, they yelled in horror as they woke up. But soon Tioti began to laugh. What sounded like skeletons falling apart, their dry knuckles rolling down between the boulders, were large snail shells, most of them the size of chicken eggs. They had long since been left vacant by the lazy sea snails that had manufactured them, and were carried about on the backs of busy hermit crabs. What had felt like sharp fingernails pinching me when I grabbed them in the dark were the claws of the crab, who had been trying to run away with the stolen house on its back.

We were to discover that we had fallen asleep in a place where thousands upon thousands of hermit crabs rattled about at night. Even Tioti had never seen so many. What were they doing? The biggest were the size of a child's fist, the smallest the size of rice grains. They start their lives looking for a house to live in. The spirit of nature, instinct, that inaudible voice, tells them that they have to go and look for empty snail shells of suitable size.

As tiny youngsters, hermit crabs need a tiny shell. They

inspect old and abandoned snail shells on the beach, and each selects one and skillfully threads it on itself as an armor to protect its soft, shrimplike body. The hermit crab will eventually out-grow its first shell and must look for a bigger one. It will stop beside the little old house next to the bigger new one, as if to compare their sizes. Then the tiny crab will pull its long curvy body out of one shell and quickly put it into the new, moving happily about with the new house like a tent on its back until it is time to move into a still-bigger shell. The long soft body of a hermit crab has been given a spiral shape that enables it to fit the curves of the snail shell, and one claw is large, hard, and so precisely shaped that it serves as a close-fitting lid or door to the stolen house.

All the hermit crabs in the world go to look for abandoned snail shells to protect their soft bodies. The octopus also has a soft body, but none of that species imitate the hermit crabs. Why? Does the hermit crab think individually, or are all of them guided by a common remote control on a frequency missed by the octopus? The answer is instinct. The magic word devised to camouflage our ignorance.

There is another kind of crab that is told by instinct to do something different but equally smart. The spider crabs pick young seaweed from the rocks with their claws, and with the skill of a gardener transplant it onto their own backs. The purpose of this is camouflage. As the seaweed grows, the spider crabs can move about undetected by predators, hidden by the garlands of seaweed that move above them.

Instinct has given a different idea to the dromia crab. This distant cousin of the hermit crab putters about below water looking for a certain kind of live sponge. The sponge is inedible, but that is the point. The crab picks it up and transplants it onto its own dorsal shell. The potato-shaped sponge grows to become as big or bigger than the crab, but is extremely light and thus no burden. The smart fellow underneath the sponge crawls about on the ocean floor, always completely hidden from crab-eating predators. The obvious reasoning behind this idea is that the sponges are not edible, while the cautious crab is. Once again a smart idea had been programmed by inventive instinct.

I had long since begun to feel that biologists and laymen alike had taken the ever-present and never-missing force we call instinct too lightly. Behind it was hidden the key to untold treasures of scientific knowledge and even to religious understanding.

On our way back to Omoa Valley we had an adventure in Tioti's dugout canoe that did not cure my childhood fear of deep water. We lost some time because we had to load the canoe with fruit and Tioti was hauling in fish. Before we knew it the tropical sun sank vertically into the sea, leaving us in sudden darkness. Occasionally we could see stars, as huge clouds slid past and left us for a moment with an obscure impression of Fatu-Hiva's skyline. But for most of the time Tioti steered by the waves ... and we saw no sign of land.

The waves grew higher, the long swells got sharper backs and some began to hiss with patches of foam that flashed in the dark. The little canoe would rush downhill and quickly start another climb, its shape making it unable to tolerate rolling. We could not see the waves in time to prepare. We threw the fruit and the fish overboard. All three of us had to bail each time a treacherous crest suddenly poured into our laps. We bailed for our lives, I with a big gourd bowl, Liv with a coconut shell, and Tioti with his hat. It was crazy to pour fish blood overboard to the sharks, but the hull had to be emptied before the next wave broke in. Two drenchings would have filled us to the brim.

If the water level inside and outside the hull were permitted to meet, it would be useless to bail any more, Tioti remarked. I swore I would never venture into the ocean in an open canoe again. Not that the wooden canoe would sink, Tioti explained. We could keep on paddling with our heads above water – if it were not for the sharks. These were the world's most shark-infested waters, because of all the goats that fell down from the coastal cliffs. Sharks have special organs that enable them to smell with their whole bodies, and they can detect a single drop of blood from incredible distances. The largest blue sharks in the world patroled the cliffs of the Marquesas Islands, some double the size of a canoe.

I began to fear that we would never float long enough to reach Omoa, hidden somewhere in there in the blackness. We paddled and bailed. We mostly bailed. What a crazy idea to travel on the ocean in a bathtub that can be filled by waves. The Polynesian outrigger gave us an amazing stability, however. It seemed to make smooth the crest of the waves. Little did I suspect that ten years later I was to enter voluntarily into this same vast ocean on nine logs lashed together in a way that enabled the water to run out between them. Nobody could have convinced me, while seated in Tioti's canoe, that I would come to love the endless sea and the beauty of its dancing waves.

That night off the cliffs of Fatu-Hiva I fully realized that we were on a tight rope, balancing between life and death. I felt I was on the threshold of getting an answer to the argument between my parents as to whether there was a life after death. I felt there was. I took comfort in thinking of my father's bedside lessons. I began to paddle with added strength. Tioti noticed, and used more powerful strokes, too.

At last we came close enough to the cliffs to recognize the familiar rocks at the entrance to Omoa Bay. A huge campfire had been lit to show us the way. Pakeekee was expecting us. We rode in on the wild surf and were grabbed by strong arms that dragged us up to the campfire. I was longing to be inland, far away from the surf and the sea. Happy to be alive, we preferred a thousand mosquito stings to one shark bite.

Down in the village, everybody still waited for the schooner. But no schooner came. Weeks had passed. Months. When the third month passed and the island lookout had seen no sign of sails or smoke on the horizon, the conditions in the two island villages became unbearable. And worst for Père Victorin. Now he wanted to escape from the island at any cost.

IX

Exodus

THE LARGEST VESSEL on Fatu-Hiva was an old discarded lifeboat once used for loading sacks of copra aboard schooners. Willy had abandoned it as useless due to its cracks and rotten planks, and it lay under some palm leaves in the grass above the boulder beach. Père Victorin had staggered down in his long black robe to take a look at it as if he were planning an escape.

Not even the biggest dugout canoe on the island was considered safe for a trip to the land hidden beyond the horizon. Tahuata and Hivaoa were the nearest islands. Although they were of volcanic origin like Fatu-Hiva and their skyhigh peaks were tall enough to arrest the trade-wind clouds, they could not be seen from sea level. But we had admired their ragged crests silhouetted like castle ruins on the far horizon when we were up on the highland plateau.

This was the worst season, and the sea was as angry as a shoal of fighting sharks, its whitecaps flashing in endless rows like snapping teeth. But the strong east wind had veered and was blowing more from a southeasterly direction ... favorable for a northbound trip toward Tahuata and Hivaoa.

No news from the outside world reached Fatu-Hiva. The island had no wireless station. Nobody had a radio.

When six months had passed since we last heard news, Willy was sure that something was wrong. When World War I raged from 1914 to 1918, Fatu-Hiva had lived in happy ignorance about the fighting until it had ended. In Tahiti, Papeete's wooden houses had been shot to splinters by a German battleship. But nobody had brought that news to Fatu-Hiva.

I began to fear, like Willy and Père Victorin, that the outside world was in flames. When we had left Europe, a terrible civil war was raging in Spain. Another was being fought in China. Modern man had learned nothing from World War I. Nothing but how to make much more effective and terrible arms. Never in human history had so much money and scientific effort been spent on inventing ingenious means of manslaughter. What progress, to fight with tanks rather than with clubs like the former cannibals on Fatu-Hiva! Now they seemed like gentlemen's tools for fair combat. The spectacular progress of the modern world had bypassed any ethical standard and Darwin's theory of evolution was not confirmed by either muscular or mental improvements in man since the Stone Age. I had come to Fatu-Hiva convinced that a second world war was bound to occur, because modern man had gained no wisdom from the first. Nothing from history and all the great cultures that had collapsed before ours got its turn to try survival.

I was impressed when I learned what Père Victorin was up to. On instructions from him, a few of the islanders dragged the discarded lifeboat out from the palm leaves and began patching up the hull by replacing the most rotten parts with new bits of wood.

Then the shabby little craft was pulled out into the bay and left afloat, tied to a stone anchor, to allow the planks to swell. A few slender trees were felled near the village, and from these the men made a mast and rigging that seemed fit for a vessel far bigger than a little open lifeboat.

At last, on a morning as cheerless as all the rest, Père Victorin made his decision. With a chosen team of native rowers, he was conveyed by canoe to the anchored lifeboat, and a few moments later we all saw the fragile craft heading into the vast ocean. The black figure of the little Frenchman sat motionless amid his crew. We had to admire his courage. The frail boat rose and fell on the huge swells and shot ahead as soon as the big sail was hoisted. Sail and all else were swallowed up completely at intervals by mighty jaws of water, and we held our breath in fear that the departing group would never emerge again. But a moment later they would ride high on top of the next towering swell. They emerged and vanished and showed up again, until the white sail became a tiny speck and was lost from sight entirely. We shivered. This could never end well.

Exciting days were to follow. The villagers believed that little Père Victorin with the big legs would land on Hivaoa, the main island in the southern part of the Marquesas group. His Fatu-Hiva crew would come back to the Omoa village with rice, flour, and sugar. For a full week, expectations dominated the feelings in the village.

With Père Victorin gone, we were greeted by the islanders with the same open friendliness as before, and Pakeekee and Tioti were again accepted by the community. They seemed to be seated together down on the boulder beach most of the time, watching the northern horizon.

Then a lookout at the cape signaled that the lifeboat was in sight.

On the verge of complete exhaustion, the men and their open boat were washed ashore by the roaring surf. They had rowed back across the sea. The mast was broken and gone. A bottom plank had been smashed in, and all the rowers had bailed for their lives while the hole was patched. The priest had been set safely ashore on Hivaoa, but a tiny sack of soaking-wet flour was all they had managed to rescue from the billows. Dripping with seawater, each man staggered to his own hut, and none opened an eye for twenty-four hours.

For all of us now, conditions grew even worse. No schooner arrived. The radio operator on Hivaoa had said that there was no world war, only copra speculation.

Now that the natives were more friendly again, we came down for seafood and began to feel less fearful of disease here than at our insect-ridden jungle home. In the village area down by the beach, there was a constant sea breeze, which drove away most of the mosquitoes. But the stagnant water lying on the beach was obviously filled with invisible germs from the village that polluted the outlet of the river. The slightest scratch of our feet resulted in ugly *fe-fe* wounds.

One day when we went down to the beach to solicit any possible news, we were invited by Pakeekee and Tioti to join a very special fishing enterprise. We were to fish in the air, not the sea, Tioti explained.

A moonless night. We paddled outside the bay in two canoes. The sea was calm, and each canoe had a torch of long *teita* grass burning in the bow. Snapping and crackling, they flamed up and lit the surrounding water while a rain of sparks danced into the night air.

Suddenly, fish began to sail about us. They shot out of the night water like big, glittering projectiles and passed through the torchlight before diving back into the black sea on the other side. It was flying fish we were out to catch, but not with a hook and line, and not with a fishing spear. We were to catch them in the air like birds.

After a little while, we found ourselves at the center of a very lively scene. The sparkling torches had an amazing effect. They brought the flying fish out in great numbers. The streamlined fish came gliding through the air, attracted by the flickering light. We had never seen such big and heavy specimens anywhere. They were the size of our forearms. They sailed through the air with the speed of an arrow from the bow, and some of them struck the side of the canoe with a heavy thud.

'Mind your eyes,' warned the sexton. The glittering projectiles seemed to lack any steering control as they whizzed past our faces.

The sexton had risen to his feet and stood balancing in the narrow canoe, fencing in the air with a net on a bamboo pole. Each time he caught a fish he quickly reversed his net to dump the catch into the canoe. There the flying fish was as helplessly grounded as a glider. In order to fly, this fish has to work up a terrific speed underwater before it can leave the surface and glide through the air for a hundred yards or more by spreading out its oversized pectoral fins. How perfectly streamlined they were for speed in contrast to the clumsy airplanes of the time! One hit me in the stomach with such unexpected force that I fell off the loose thwart, to everybody's wild amusement. Tioti got one that sent his hat dancing overboard. We ducked and twisted to escape, and Tioti fought and fenced with his net to catch the projectiles sailing in all directions through the arena of light. This was the kind of fun that made Liv and me recollect the snowball fights of our happy childhood in wintry Norway.

As the torches burnt out and we returned to shore, I counted thirty-five large flying fish in the bottom of our canoe, of which quite a number had arrived without benefit of Tioti's flying net.

We were sitting one day with a group on the stone ledge around the huge banyan tree down by the beach. Still no boat. With

Willy, I was making a decision. We had no choice. The naked flesh in Liv's wounds had begun to look like speckled salami. She was able to take it calmly, but I was not. We had come to a dead end. We had to try the same trip as Père Victorin, away from the island. Willy and several other men from the village were planning another voyage in an attempt to bring back rice and flour, and we decided to go with them.

The bamboo cabin was abandoned. We would have to return for all our zoological and archaeological treasures, but how and when we could not tell. Meanwhile, we were worried that they might be stolen. Our bottles of bugs and other creatures would surely interest no one. But we had obtained by trading some archaeological treasures that any thief could resell. Theft was not a very serious sin to these people, and the Christian God was merciful. But they feared the Devil and the magics of taboo.

I decided to try my luck as a medicine man. Some of the less-trustworthy youngsters from the village followed us as we went on our last visit to the cabin. I opened a jar of formaldehyde, used for the preservation of zoological specimens. In turn, the young men sniffed eagerly at the bottle, then, with burning nostrils, began to leap about, grimacing and sniffing. I turned over stones until I found one of the large centipedes of the poisonous kind that would crawl about even if cut into segments. If dropped into water, it would wriggle about like an eel. I dropped it into a jar of formaldehyde, and the eyes of the spectators grew big as they saw the creature become stiff and dead in a moment. Now I splashed some of the liquid upon the bamboo floor and jumped out, shutting the door with a bang, bolting it with poles and cord. In a moment, the deadly vapor would fill the room, we explained, and until we came back, no one could enter without suffering the fate of the centipede.

That night we went to bed on the floor of Willy's house. We were to leave the island at a very early hour.

*

It was still night when we were awakened by dogs barking and saw the lights from lanterns coming from the forest. Noisy groups of strong men gathered outside the door. Everything had to be prepared for an early departure. They all gazed at the weather. The clouds sailed wildly across the night sky. Surely the sea was rough out there. Better to wait and see.

It was cold at that hour. Willy put a kettle on the wood fire. Orange tea warmed our sleepy bodies. We felt more awake. It was strange to sit among former antagonists. Neither Tioti nor Pakeekee were there, but Ioane was sipping his tea from a bowl and throwing quick side glances at the weather. There was a light drizzle.

The rumble of the surf seemed particularly sinister that night. It had never before affected us this way. We shivered from drowsy chill and suppressed fear. One or two of the men tried joking, with chattering teeth. The rest remained silent.

It seemed to get slightly lighter, as if dawn were on its way, the very first indication that the black night was loosening its grip. Willy rose and gave an abrupt order: 'Hamai!' (Come!)

This was it. We were about to embark on a lunatic sea voyage. There was nothing I dreaded more than another confrontation with the sea, but it would be still worse to remain behind. With fresh banana leaves wrapped around our legs, we staggered after the others, waddling down to the boulder beach, where there was no more time to think: our ears and minds were filled with nothing but the rumbling of the thundering surf on the slippery boulders.

Experienced paddlers took us out by canoe to the dancing old lifeboat, now repaired, and anchored at a safe distance from the rocks. Then the canoe returned to bring the second group through the witch's caldron. As the canoe came out for the third and last time, one person remained on the beach, wildly and angrily waving his arms: the little Chinese, who had come from Hanavave and also wanted to escape. But there was no more room – and the poor chap had even brought with him a large pig and a flock of live chickens which he wanted to take on board! Our last picture of Omoa was of that tragicomic man standing on the beach with his pig and his fowls, desperately waving to be picked up. But the lifeboat was already loaded far more than made sense. There was frighteningly little of it above the water, and we were still only in the sheltered bay. The scanty freeboard would have made the enterprise seem quite crazy but for the fact that our brown companions had made the voyage before with Père Victorin.

Quantities of *fei* and bananas lay stacked up under the thwarts, on which the crew doubled up in pairs, and a barrel of water was stored in the bow together with green coconuts. If all went well, with such a wind the voyage would be accomplished in a single day. But we had neither compass nor chart to give us a course. We could only steer into the empty ocean in the general direction of the other islands, and stick to one course until the summits of Hivaoa rose above the sea. If we ran into fog, we had no chance unless the weather cleared before we drifted away from the whole island group.

Fully laden, the old lifeboat rolled perilously in the swells. The Polynesian crew took a last look at the weather. The strong southeast trade wind was at our backs.

We were thirteen. Old Ioane was the captain and sat high astern with an oar serving as a rudder. At his feet, in the bottom of the canoe, lay Willy, Liv, and I, leaning against a pile of shabby suit-cases and sacks. In pairs on the four thwarts in front of us sat eight rowers waiting, the eight best from the village. With deeply marked, almost brutal faces, which seemed to defy any kind of weather, and with muscles rippling under their coconut-oiled skins, they sat ready. One man was in reserve, with a big, wooden bailer. A hammer and some rusty nails lay ready beside him, in case a plank should come loose when the waves struck the bottom.

Ioane sat waiting impatiently, dressed as usual in a straw hat, white shorts, and undershirt. He looked as if chiseled from stone, his wrinkled face grinning toward the drifting clouds as he studied the course of the rolling waves.

All was ready. Ioane rose to his feet and uncovered his head as he crossed himself, touching first his forehead and then his chest. Tough and serious, the rest sat with bent heads while Ioane slowly said the sailors' prayer in Polynesian. When he finished, everybody crossed himself and the ceremony was over.

It was as if a thunderstorm had suddenly hit us. Ioane swung his arms and shouted his orders. The others leaped from the thwarts, yelled and bawled; the stone anchor was hauled on board and the huge sail was hoisted with terrific hullabaloo, while the whole vessel threatened to capsize. All aboard went completely wild. Even the taciturn Willy began shouting and giving orders.

Like a bird on the wing, we shot across the waves. The natives beamed excitedly and shouted wildly as the large Polynesian sail was properly lashed and caught the wind fully. This was the life of their ancestors, something that made the blood rise in this drowsy, sleepy, modern tribe. They cheered and rejoiced while Ioane sat grinning, hunched over the steering oar, his unshaven face bristling and beaming with the joy of living. This was fast going, and for two tired jungle dwellers, too, something to make the blood course through the veins.

But we were not to remain long in the shelter of Fatu-Hiva. As the lofty island slid astern, the wave valleys got deeper and the swells of the open ocean rose above us, while Fatu-Hiva shriveled to a mere rock in the frothing sea. One moment the wave crest rose to fumble toward our open vessel, next we were whipped up on top, the lukewarm sea drenching our bodies and slapping salt sea spray against our faces until we could hardly see for the burn of salt and sun. No sooner was the frail boat at the top than we looked down into another green and glassy valley, followed by another white and hissing crest. And then we were suddenly once more deep down, looking up again in awe at the water tumbling high above. This was sea wild enough to toss about a vessel much larger than ours. We were in fact to learn later that at that very moment the schooner *Tereora* was pitching northward somewhere near us, in the same seas. Captain Brander reported that the sea had been so agitated that his schooner was battered as water crashed over the whole vessel, burst through the galley door and caused havoc and destruction to his cargo below. The sea had been too choppy for the big schooner to pitch freely full length between the waves, while our little lifeboat had more room for dancing in harmony with each single wave. With our large sails as wings we flew up onto the wave crests and rode the heaving water masses in a dizzy race.

At the steering oar, Ioane was incredible: crouching and grinning, he was on the alert for every chasing comber that threatened us, and he rode it off. If it burst and sent the breaking surf crashing into the whole vessel, he clung to his steering oar with almost diabolical determination, and not for a moment did he let his attention wander from the next wave that rose up in pursuit. While seawater streamed down his face and salt burned his eyes, he clung tightly to his oar and was on his guard. He was a truly magnificent captain.

Two of the younger men soon lay huddled in the water at the bottom of the boat while the others growled ironically and made jokes about their seasick companions. Drenched in seawater and with the banana-leaf bandages long since washed overboard, Liv began to look terrible, her legs grossly swollen. Raw flesh was literally bursting from her wounds. At the height of the day she passed out, and lay lifeless against our bag while I held her tight each time the seas broke over us.

Several times, the boat was almost filled with gushing water, and it was as if we were sinking. But soon it rose once more, the bundles of bananas floating in the seawater while all the rowers bailed for their lives. The excitement was hectic, without a moment of repose. Time and again I abandoned all hope, as we raced along the wall of a breaking swell that rose on end and tumbled down upon us in massive cascades of water before we managed to sail clear, or when we raced crazily over a tall crest and took a leap down into the valley so hard that the planks creaked and columns of water tumbled away on all sides. I was also increasingly worried about Liv, who lay against my legs with her eyes closed, not even reacting when heavy cascades of water pounded over us.

I was still a sworn landlubber, but I began to digest a few lessons from the sea. This was my second experience in a small craft on the open ocean. The first had been in Tioti's dugout canoe. Again I began to wonder why the ancient boatbuilders had stopped building buoyant log rafts and begun to produce sinkable boats with frail, planked hulls. In this lifeboat, as in the canoe, we were struggling for our lives, bailing all the time. I was clinging continuously to the hope that there was some invisible power behind the wonders of nature that could respond to my intense desire for help and mercy. How stupid to make a boat of thin planks, no more than a receptacle for the breaking seas. I wished we had been sitting on a raft that water could not fill. A raft would obviously have been quite safe between these same waves. But man had long ago changed marine architecture, preferring speed to security.

Our speed was breathtaking, but our lives were at stake. For a long, long while, we saw no land in any direction, even when we rode over the highest crests. Fato-Hiva had vanished, but Ioane steered on as confidently as if he had a compass.

Not for a moment did it occur to me that we had run into an exceptionally rough sea. I was greatly impressed by the height of the waves around us, but I thought that any ocean wave would look as high and scary when seen from such a tiny vessel. Not until we heard afterward of the seaworthy trading schooner *Tereora*'s troubles in the same area the same day, did I understand that there was something exceptional about the seas we rode. Not knowing this yet, one thing was clear to me: it was thanks to its small size that our open lifeboat kept afloat. There was room for the whole vessel between the waves; if it had been a little longer, it would have stretched from one wave to the next and either the bow or the stern would have cut into the walls of water around us. It was a landlubber's misconception that the perils at sea are greater the smaller the vessel.

I cannot remember how the hours passed. All I can recall is that a truly burning sun dazzled our eyes and struck our skins like hot sparks between the cold and heavy cloudbursts. The copper-colored backs in front of us turned dark during the trip, whereas the salt spray and the reflection from the sun on the gleaming waves made our own skin rise in blisters. Our flowing hair was our only protection from sunstroke. I recall the puffing of a shoal of glittering black porpoises that emerged on all sides, dancing about us in a wave trough, then we and the whales were all lifted up by a mountain of water and parted company in the confusion of waves as we ourselves raced on in a wilderness of water.

Farther and farther we sailed. By now, it must have been late afternoon. The distance we had covered had been increased by sailing up and down and by zigzagging between the waves. Half asleep, I heard a shout from Ioane: 'Motane!' Motane was a small, uninhabited island to the north. Ioane altered course, for we had to pass to the west of it.

New life seemed to have been injected into the blackening brown backs. The men on the thwarts, too, were sitting half drowsing, guarding the long boom of the sail to prevent it from colliding with the tumbling waves. There seemed to be more things happening now. To the far north, we kept seeing the gray table mountain of Motane rising like a whale's back whenever we traveled across a high swell. We climbed and we leaped and we skirted alongside the wave walls just as before, but now at least there was something in sight. Only Liv did not open her eyes.

After a while, the contours of Tahuata Island rose indistinctly from the sea to the left of Motane, with haze obscuring what seemed to be patches of jungle or green valleys running up the blue mountainsides. This inhabited island froze in a sort of mirage that seemed never to come nearer. There was a long way left to go, for although not quite as high as those of Fatu-Hiva, the mountains of Tahuata rose to more than three thousand feet above sea level and could be sighted from afar.

Like a thin veil, Hivaoa at last emerged to the far north, outlined as a long and grayish-green ridge between the other two islands, but much farther behind. I shouted to Liv that we could see our destination, but she heard nothing.

The day went on. The most dangerous stretch of water now lay before us. Our traveling companions all knew that a strong ocean current forced its way westward through the narrow passage between Tahuata and Hivaoa. The wide westbound current was forced at increased speed into a sort of huge funnel as the first land masses since South America impeded its unhindered drift, and a mad reflex of waves and rushing water chased up angry billows far out in the ocean at each end of the channel.

We had to cross this area to reach our destination. Willy was admittedly afraid to venture into the treacherous belt in the existing weather conditions, and was deliberating with Ioane, but they could find no alternative. The best Ioane could do to reach Hivaoa was to hold as far to the east of the channel as possible. The sea would be worse the closer we kept to the coast of Tahuata. This was my second lesson from the sea. The textbooks were wrong. I had always heard from anthropologists that primitive people could travel at sea only by hugging the coasts of continents and islands. Experience showed me again that nowhere is the danger of ocean navigation in a small boat less than at the farthest possible distance from any treacherous coast.

We darted at full speed into the marine wilderness swept up by the confused current where it hit land. Ioane sat crouching astern with all senses and muscles alert, like a puma ready to spring. Everything now depended on him.

Far ahead to the left was the long headland leading to the largest valley on Hivaoa: Atuona. This was a sort of capital, at least for the southern islands in the Marquesas group, the only competitor being the Taiohae Valley, on Nuko-Hiva, to the far north. We knew that some two or three hundred Polynesians lived in Atuona Valley. And the French governor had formerly lived here. We also knew that this same valley, which now began to open up indistinctly ahead of us, had been the final home of the reknowned painter Paul Gauguin. Few people knew that he had also been a skilled wood carver. I sat in the dancing lifeboat with his gun. The wooden butt-end of the old rifle was artistically carved in typical Gauguin style into what seemed to be a deity, perhaps the god of the hunt traveling in a chariot drawn by oxen. He was seated, riding backward with a raised beaker in his hand, presumably traveling across the sky as there were undulating scrolls suggesting clouds over and around the traveler. I had bought the gun from Willy, whose father had received it as a gift from Gauguin once when he had visited Fatu-Hiva. Gauguin's tomb was in the hillocks in there somewhere.

Black clouds and cliffs covered the evening sun as we came abreast of the headland leading into Atuona Bay. The cliffs, exposed to the ceaseless east wind, blocked the freedom of the high rollers we rode, and a continuous belt of white cascades and geysers was flung skyward along the steep rock face.

Drenched, exhausted from bailing, and stiff from salt and sunburn, we came through the area of mountainous seas and prepared ourselves for a hazardous landing. The swells still rose high as we turned in close to the rocky shore, and the hoarse commands from many voices mixed with the thunder of the surf as the sail and mast were lowered.

The black sandy beach of Atuona lay in front of us; we had reached the journey's end – almost. Yet none of us felt the slightest relief, as we could see and hear the inferno that separated us from land. Wind and waves came straight in from the agitated ocean to the unsheltered beach. This time, we were not to land on the leeward side of an island, as we had on Fato-Hiva. Four thousand uninterrupted miles had brought the surf in front of us from the Pacific coast of South America to this island.

The eight strong rowers shoved their oars into the rowlocks and Ioane set the course straight for the center of the long beach. The sandy bay was incredibly shallow for hundreds of yards in front of the beach. Row upon row of towering breakers raised themselves on end with irresistible force, and although we saw only the long sequences of sloping wave backs, we could clearly visualize the glass walls of water as high as houses that rumbled and tumbled in across the shallows toward the land. The white streamers and the roar of the water left us in no doubt. And on this day the surf went beyond the sandy beach, flooding the green meadow at the foot of the coconut palms. There was still light enough to see a group of natives gathering as if to watch the mighty display of nature's forces.

Liv was fully conscious now, but she watched the surf in front of us with tired indifference, as if it did not concern us. All the men were seated, oars ready, waiting for the slightest command from Ioane. The two who had suffered from seasickness were sharing one oar in this last effort. Time and again, we rowed slowly toward the edge of the surf area, but then some disturbing wave would rise behind us, and the rowers were forced to back the oars as fast as they could, while the rising crest tipped the stern into the air like the tail of a sea bird. Then we ventured forward once more.

Ioane sat grimacing like an angry devil, yelling his orders until his voice cracked. He was watching every oar. Crouching almost double, he seemed to be alight with fight and fury. At this moment, we all looked to him as the leader. His brain served for all.

The right wave came and caught us from behind just as Ioane shouted, 'Row, row, row, row!' as if in ecstasy.

Creaking oars rumbled in the rowlocks: the men rowed desperately, clenching hands and teeth, eyes sparkling with effort and excitement.

As if we were all on one surfboard, we were carried by the racing wall of water all the way in across the shallow area, with rows of breakers thundering in front and other rows rising one after the other behind us. We were in the middle of an inferno. We grabbed the gunwale with cramped fingers, clinging to the hurtling, unsteady boat. Then the grip of the two young men with the same oar slipped. Ioane raved like a madman, and in two bounds Willy passed over us and grabbed the loose oar, while the two men sprawled on their backs on top of the fruit.

But it was too late. The lifeboat turned sideways, and Ioane had not the slightest chance of taking hold of the chasing wave with his steering oar. The crew tossed away the oars and, like acrobats, sprang onto the gunwales. Then they all threw themselves overboard and disappeared into the sea.

I grabbed hold of Liv, and together with Willy we jumped away as the boat rose on end and turned bottom up.

It was due to no effort of my own that I rotated, helpless as a starfish, in whirling water until I realized the sea itself had delivered me seated next to Liv in the shallows. A pursuing glass wall was on its way in to get us, and we felt the suction as the backwash came like a river, pulling us seaward. But, clutching hands, we fought our way up onto the foaming beach and the safe grass beyond.

We noticed our traveling companions clinging like ants to the dancing lifeboat in the wild surf, which had rolled over time and again. They did not want to lose it at any cost, and while boat and men completely and repeatedly disappeared in the frothing water, not one of them let go. Then, more under the water than above it, they came swimming ashore with sacks and suitcases.

Finally the boat, filled to the brim with water flowing in and out, was tilted up and pulled out of the ocean. To put it beyond further risk, the old and battered lifeboat was carried up onto the turf and well inland between the palms.

The oars, the fruit, some straw hats, and other trifles were politely delivered by the sea and saved.

We had landed on Hivaoa.

X

Stone Men Against Fossil Dogma

What a beautiful valley. Freed from the horrible vision of ending our lives in a salty graveyard, we wiped the brine from our eyes and experienced another vision of Paradise: Atuona, a world of its own, surrounded by a sky-piercing wall of precipices, shut the rest of the world out on all sides but to the south, where the horseshoe shaped mountain range opened toward the blue ocean.

What a lovely world of its own, indeed.

It must have been a dream place for an artist with the eye and feeling for shapes and colors like Paul Gauguin. The valley was the same as in his day, the same warm composition of simultaneous delight that filled all the senses. Colors that one could almost smell and taste as well as see. A few steps on warm black lava sand, and we walked onto a thick carpet of juicy grass. A footpath through low bushes of red hibiscus flowers led us away from the beach, and in front of us lay the valley itself, guarded by an army of tall coconut palms standing at ceremonial attention. They posed majestically with long necks and large, ornamental feather headdresses waving high above the tropical exuberance below. Amassed around their legs were still the subjects so beloved of Gauguin: oversized tropic leaves, yellow fruits, red flowers. And, as in the days of that great painter, a few modest huts of imported planks lay scattered in between the splendid compositions of nature.

Atuona Valley on Hivaoa was the main port of call for the occasional yacht that cruised through this remote part of the Pacific before World War II. Even here, however, the number of calls from the outside world were exceedingly few. The Marquesas group lay outside shipping lanes, and globe-trotting yachts were discouraged by the lack of a harbor and the need of government permission to remain ashore for more than twenty-four hours.

But as we got to our feet and looked around, we could see from the total disinterest of the crowd that had witnessed our unconventional arrival that we were by no means the first Europeans to visit this bay, although we might have been the first white couple to arrive upside down. The little group of Polynesians of mixed blood that watched us looked with contempt at the few and shabby bags of drenched cargo that came ashore, then studied Liv and me with the eyes of experts. To them, there were three categories of white foreigners: officials in uniform, whom they feared but admired; tourists, whom they laughed at but welcomed; and copra workers, whom they despised.

To them a uniform meant power. Inside was one who made laws and could send people to prison in Tahiti.

A tourist was respected because of his wealth, but he was an ignorant fool. He could not tell the difference between a *fei* and a banana and did not know when the rainy season would start.

Although he was Christian he bought old pagan *Tiki* images. And he paid more for them the older and more shabby they looked. So, to get the best price, the islanders carved fine new ones and buried them for some time to make them look miserable.

But worst was the copra worker. He was as white as a tourist but had no money and worked as an islander. He had more brains than the tourists as he did not ask all those silly questions, and some could even climb a coconut palm or get as drunk as any islander. But he always came to gain and not to give. Besides, tourists and government officials did not seek his company, so he must belong to an inferior kind.

As we staggered up the trail, drenched by the sea and scorched by the sun, with our legs in a miserable condition and with no other earthly property than a rusty gun and a shabby, water-dripping bag, there was no doubt in the minds of the spectators: we belonged to the third category.

Willy and Ioane had local friends and were immediately taken care of. One by one our other Polynesian companions from the boat also disappeared into various houses in the village. Left alone, we headed for the French *gendarmerie* and knocked at the door for advice.

A skinny white man in uniform opened the door and glanced at us apathetically. Keeping his tropical helmet on his head and his right hand in his trouser pocket, he stretched out his left hand to Liv. Then he caught sight of the old gun in my hand and asked to see my arms license.

I explained it was an old souvenir from Paul Gauguin.

'It is an arm,' the gendarme repeated, and the old rifle was confiscated.

Some days later, with my legs in bandages, I came back and asked for a screwdriver. I unscrewed the wooden butt from the rest of the gun.

'Which part is the arm?' I asked, and held one piece in either hand.

The gendarme immediately pointed at the metal piece with the long barrel. I grabbed the wooden butt and left his office happily with Gauguin's wood carving in my hand.

In the meantime, Liv and I had found room and board and ate marvelously in the bungalow of Chin Loy, a Chinese who ran a tiny restaurant in his own kitchen. We had run into a marvelous French couple who came walking along the village road, loaded with cameras. He was a photographer, she a journalist, and they had come with the weather-battered *Tereora* the same day. The schooner still lay at anchor in the next bay, which was better protected than the open beach where we had made our landfall.

Before they helped us to find a place to rest, they had discovered the horrible wounds on our legs. We staggered with our new guardian angels as fast as we could to the only bamboo cabin in the valley. This was the bright and surprisingly pleasant island hospital. It was run by a most winning and cordial young male nurse from Tahiti, named Terai, 'Heaven', the name Teriieroo had given to me, also. My namesake actually reminded me of Teriieroo, in both body and mind. We immediately became friends.

Only in his twenties, of medium stature and yet with a massive 250 pounds crammed into his brown skin, the crew-cut Terai was a fervent sportsman and incredibly agile. He was born in Tahiti and had spent some years assisting at the hospital in Papeete, where he must have used his time well. With a mere glance at our troubled legs, he identified our wounds as tropical ulcers. If we had come a couple of weeks later, he said, Liv's infection would have reached her bones, and she would have lost a leg by the time he could have arranged her transfer to the hospital in Tahiti. This proved to be the sad price paid by the Omoa woman who had refused to follow Père Victorin or ourselves on the lifeboat crossing to Hivaoa.

Surrounded by a group of islanders suffering from ailments ranging from toothache and venereal problems to the recent loss of a finger, we took turns to stretch out on a bench while the corpulent Terai selected tools from a box of knives and pincers.

A week later, we could look back on the first visit to Terai's bamboo hospital without feeling pain flashing from head to feet. Terai had used his instruments well. He had cut and poked, he had pulled out toenails to stop infection from reaching the bones, and he had smeared on thick coats of a magnificent yellow-green ointment scooped out of a big pot. Soon, we were feeling better.

At the end of the week we learned that the *Tereora* was ready to weigh anchor, calling at Fatu-Hiva on its way back to Tahiti. But Terai would not permit us to return to our own island, for he insisted that our feet were still in danger unless we continued his cure. We staggered to the rocky cape to get a glimpse of the *Tereora* and to wave to Captain Brander, who never came ashore. Farewell to all our friends. Vivacity and energy seemed to shine from the bushy red head of the petite French lady as she shouted a last *au revoir.* Holding the little, camera-loaded photographer by the hand, she jumped with him from the cliffside into *Tereora*'s dancing lifeboat, both of them falling like drunkards into the waiting arms of experienced brown seamen.

Then the *Tereora* hoisted her canvas and sailed away. On her deck we saw Willy, Ioane, and all our other friends from Fatu-Hiva well provided with sacks of food.

Our thoughts went ahead of them, and for a moment, once again, we were sitting looking out at the splendid view from the window of our bamboo cabin. Then we shook the memories away as we seemed to inhale the scent of bamboo dust and feel our bodies itching. We returned to Terai's bamboo house-hospital to have our endless yards of bandages changed.

Not much happened during the weeks that followed. We staggered back and forth between Terai's house and the screened compartment of Chin Loy's exotic kitchen, where we drank in the gentle atmosphere of age-old Chinese culture and ate our fill of every delicacy a Chinese master could concoct by combining tins of corned beef with the rich products of the island soil.

Because of the great distances and the lack of proper water transport, Terai was never able to visit other islands in the group, but once a month he saddled his horse and rode on an inspection tour to other mountain-girt valleys of his own island. In spite of his corpulence, he was a born horseman, and his tiny Marquesan stallion galloped as if it were carrying nothing but an empty sack tied to its back.

Terai never walked a step. He kept his horse tied to his bamboo wall when not required, and when he visited the sick, with nothing but a sack for a saddle, he dashed along all the village trails. When the day came to start his monthly expedition around the island, two extra horses with handcarved wooden saddles were lined up beside his own. With much persuasion, we had got his consent to join him on the trip. Our legs were improving, and in his company we could better continue attending to the wounds. Besides, he was to visit the lonely Puamau Valley at the far eastern extremity of the island, where we had heard a solitary Norwegian hermit was living. He had lived there for many years, and had certainly not seen any countrymen since he ran away from his ship as a young boy.

Hours before the sun rose, we started the long climb to the winding mountain ridges that were to take us to the remote Puamau Valley. Once more, we felt the surging of happiness at entering virgin wilderness, while we filled our lungs with clean, cool mountain air. Large palm valleys opened below us. Green, jungle-wrapped pyramids of mighty dimensions rose into sight in the island's interior, united by ridges as sharp as horses' necks. The winding trail led up to these inland ridges, because the wild coast permitted no passage. As on Fatu-Hiva, the entire shoreline was eaten away by the ferocious ocean, which had transformed volcanic slopes into vertical walls, leaving only canyons and deep sections of crater valleys to open seaward between the overhanging cliffs. White birds sailed deep below us, outlined against the blue sea and sky, and a menacing white snake, the unbroken surf, wound along the whole coast, marking the boundary between the little island and the ocean. Wild cocks crowed in the valleys where the sun had not yet risen, and others answered high above in the sunlight. The horses neighed merrily, their unshod hoofs trampling upward and ever upward along the red trail.

When we reached the central crest, we stopped, having suddenly found ourselves on top of everything hands or hoofs could climb. There was nothing more than sky above us. The trade wind tore at our hair and at the manes of the horses, and the beasts reared restlessly. We began to scan the now far more distant horizon for a glimpse of Fatu-Hiva. Tahuata was wrapped up in thick cloud banks, and other clouds hung at intervals and cast black shadows on the blue ocean. We were so high up that vast expanses of hidden ocean had risen into view. There, to the south, blurred by haze where the world ended, were the rugged contours of a very small island. Only the upper crests emerged indistinctly above the water; it looked like the unrigged wreck of a burned ship sinking below its own thick cloud of smoke. It was still raining down there on Fatu-Hiva. Fatu-Hiva was far, far away.

The world of Ioane, Tioti, and Pakeekee was tiny when seen from a distance. With the eyes of the universe, mankind is reduced to microscopic proportions, and our quarrels assume absurd dimensions, like the invisible trifles we fight about.

'*C'est joli*,' said a voice behind me. It was Terai sitting on his rearing stallion and gazing over the valleys below.

'What is beautiful?' I asked, surprised, and looked at my Tahitian companion.

'The mountains, the jungle, all of it. I realize that nature is beautiful.'

Terai surprised me. He really did resemble Teriieroo.

'But isn't Papeete the ideal of beauty for all islanders?' I asked.

Terai spurred his horse. 'Not for everybody. Some of us understand better. Tahiti was good enough the way our ancestors kept it.'

We rode side by side along the windswept ridge.

'But don't most Polynesians move to Papeete if they can?'

Terai agreed. It was the tragic fate of his people, he said. All young girls are drawn to Papeete for the white men's wealth. The island boys then follow, to get their share of the amusement.

We reached a handsome mountain forest, where the horses trotted under the trees in a row, along a soft, grassy trail. Terai started singing the familiar tune composed by the old Tahitian king: 'I am happy, *tiare* flower from Tahiti.' The sun flickered between the crowns of the trees, and we had to duck to avoid the lianas and low branches that hung across the trail as the horses began to gallop on the soft carpet. Rare birds, some of them splendidly colored, fluttered or ran among ferns and foliage. We had the rare luck of seeing one of the fast-running wingless birds well known to the natives but unknown to ornithologists. Terai suddenly halted his stallion and pointed to the trail in front of us. I barely caught sight of a bird looking like a little hen without wings, when it began to run with lightning speed along the trail and disappeared in a sort of tunnel among dense ferns. None have ever been caught. How had this wingless bird landed in the hills of Hivaoa? Wingless birds in the Pacific are otherwise known only on New Zealand, where the kiwi still survives but where the twelve-foot-tall giant *moa* bird was totally exterminated by man.

Half the island lay behind us as we paused for a picnic lunch by a spring before riding into really wild terrain. The forest suddenly came to an end, and as we emerged from the trees, we felt as if we were heading into empty space. No more foliage, no more earth before us, only a head-spinning precipice. We heard a strange, distant sibilation from the surf far below, and the deep rumble of the wind as it rose up the huge wall, threatening to seize us.

As we followed Terai, who guided his horse onto a narrow shelf cut into the cliffside by the islanders long ago, we saw the world raised on end. Dizzy, we turned our heads to face the rock wall while our horses slowly, very slowly, followed behind the stout horseman before us. His mighty body seemed to fill all available space above the trail, like the rump of the horse beneath him.

Suddenly the shelf ended as the trail turned sharply to the left through a short pass that brought us out on the other side of the ridge; we faced a new precipice, this time on our left. Here, too, air currents roared up from below. The island was indeed not very wide at this point: another bay with white surf lay below us, with a drop as horrifying as the one we had left on our other side. Better to turn our noses the other way now, facing the wall on our right while the horses moved unperturbed.

Then the rock wall opened on both sides. It was like riding the winged Pegasus, with no visible foothold below us. There was a mountain peak in front of us and another behind. Between them passed an edge so sharp that the trail filled it completely. Terai turned his head, giggling. On either side, our feet dangled above steep slopes that ran down into nothing except the bands of surf far below. The surf seemed to advance in slow motion and was inaudible from this height; all we could hear was a general hiss carried up from the depths. Even the horses seemed nervous at this passage. With heads raised and ears pricked, they proceeded slowly along the ridge. The irregular blasts of wind from below seemed to scare them. I held my breath, for there was nothing to jump down upon should the horses stumble.

We were across. The trail looped around the peak in front, crossed another narrow edge, and then we rode into trees and were at once engulfed in dense jungle. When we next emerged from the foliage, we had reached the rim above the Puamau

Valley. One misstep and we would have arrived at our destination in a matter of seconds, falling three quarters of a mile through the air. We were on the edge of the mightiest crater valley we had seen on the island.

The sun sank quickly as we started our descent. Soon, only the slanting of the horses' backs, rump up and head down, told us that we were descending. Terai seemed in good shape, but Liv and I were so tired, so painfully stiff in our sore legs, so sore behind, that we did not care where the horses took us as long as we could get off the wooden saddles. It was now too shadowy to see either the hoofs of my horse or the edge of the sandy shelf. We lost sight of each other, but kept together by frequent shouts which carried strangely into the emptiness above the deep valley.

At long last, the slanting backs of the horses leveled out again and the animals began to jog-trot on grass. Next we heard the splashing of hoofs in water, accompanied by the sound of a gushing stream. Clearly, we were in the valley. The surf, too, began a rhythmic rumble – and it came, blessedly, from our own level.

A spot of light pierced our blindness and began to dance and to grow bigger between the horses' ears till it took the shape of a window. The thunder from the surf was close beside us, and a fresh sea breeze suddenly struck our faces. A house lay near the beach. Our destination. We were scarcely able to alight from the horses and tie them to some trees. The door. A gorgeous smell of fried eggs filtered through it. I knocked and listened.

The door was brusquely opened and a muscular but not big middle-aged man of Nordic appearance held a kerosene lamp above us. He scrutinized his uninvited guests with sharp blue eyes. Months, even years, passed between the occasions when this man received white visitors. And when they came, they came from the beach.

'*Bonjour*,' he said quickly.

'*God-dag*, Henry Lie,' I answered in Norwegian, saluting him in the language I knew he had once spoken.

The man backed, totally bewildered; then he recognized Terai behind us.

Yes, indeed. More fried eggs were cooked over the fire while the corks popped that night in the lonely Norwegian cabin in the Puamău Valley on Hivaoa.

Henry Lie had lived a strange life. He had come to the Marquesas group as a boy, thirty years earlier. He was then a deckhand on an old sailing vessel. The captain was a drunkard and there were daily fights and trouble on board. As the ship anchored off Hivaoa, young Henry was sent ashore with some of the crew to fetch water. He managed to run away and hide in a cave. He did not come out again until the angry captain had given up the search and the ship sailed away from the island. He fell in love with a Polynesian girl and married her, and when she became the heir to a local valley, he started a copra plantation. His wife died but left him with a son; he moved to the large Puamau Valley and built up the finest copra plantation in the entire Marquesas group.

Henry Lie saw nothing of the outside world except the trading schooner from Tahiti whenever it called to fetch his copra. His life, apart from work, was his fine son, Aletti, and his books, of which I was surprised to see he had an impressive collection. His big, one-room bungalow was filled with beds and books, witnesses of hospitality and an unexpected intellect. I was puzzled to see that he had so many books that stacks of them had to be lifted away before he could offer us three empty beds.

As on Fatu-Hiva, the entire native community lived down by the sea, where the ocean breeze drove away most of the mosquitoes. There were few houses left compared to the quite considerable population this large valley must have had in pre-European time. The handful of Polynesians living there today seemed to pass most of the time squatting in front of their huts or resting on floor mats inside. They waited for the coconuts to fall down, explained Aletti, then split them with axes and sun-dried the kernels, using the resulting copra to barter rice and canned goods from his father, who had a tiny store, like Willy's on Fatu-Hiva.

There was only one other white man living in the Puamau Valley, and he was Henry Lie's nearest neighbor and closest friend. Their common interest in books drew them together in the evening. Whereas Lie was up and around doing physical work on his large plantation every day, and thus was suntanned, his neighbor never seemed to enjoy more exercise than turning the pages in Lie's books.

Henry Lie took us over to visit his neighbor, whom he simply introduced as 'my friend'. The friend proved to be a tiny Frenchman with huge bushy eyebrows and a giant mustache hanging down on either side of his mouth. He received us in a minibungalow so small that only two of us could come inside at a time. Proud and happy, he showed us around; that is, we stopped inside the low door and stood in one spot, turning our heads. Close around us were walls patched together from empty boxes and bits of old planks and driftwood. The floor was of bamboo and the roof of bundles of grass. Never had we seen such a cramped residence with so many ingenious devices. If the old Frenchman pulled a string or turned a handle, quite unexpected things would happen. When he wanted to go to sleep, he pulled a piece of rope, and a kind of bed folded out from the wall. He pulled another rope, and a table came down. From one spot he could reach out and grasp all his accessories. If the wrong rope was pulled, a saddle might descend from the ceiling or the lid of a box might open and disclose loaves of delicious home-baked bread. He had baked them himself over an open fireplace of corrugated iron placed between the table and the bed.

He brought a warm and fragrant loaf with him, one under each arm, when Lie called us back for an evening's visit in his bungalow. Little did I expect that these two strange characters in the lonely Puamau Valley should come to have considerable influence on my life. One of them a runaway deckhand, the other a stranded jack-of-all-trades. It was doubtful that either had had more than the very minimum of basic schooling, but they could read and think, and were extraordinarily gifted in making practical observations, as I was soon to learn. But that first evening we were dead tired and crept early to bed, books piled all around us.

The sun had long been shining when we got up and out next morning. Except for the birds singing and Lie working, there was silence and no other sign of activity. As Aletti took me for a walk up the broad valley, I saw Terai's horse tied in front of one of the huts where someone lay ill. He had ordered Liv to rest indoors, so I followed my young guide inland to see the beauty of the valley.

Suddenly we saw them. The giants I had heard about. Aletti had been silent as we approached, and merely pointed as he bent the thick foliage aside. They stared at us from the thickets with round eyes as big as life belts and grotesque mouths drawn out in diabolical grins wide enough to swallow a human body. Bulkier than gorillas and nearly twice the height of a man, they had impressed the few travelers who had so far been lucky enough to visit the locality. The giants of the cliff-girt Puamau Valley displayed such a contrast to the lazy people down on the beach that the question inevitably came to mind: Who had put these red stone colossi there, and how? They must have weighed many tons.

I had read about large stone monuments in the Marquesas group, but it is one thing to read a couple of lines and another to stumble unexpectedly on large colossi in human form standing among the forest foliage.

We approached the tallest, which stood on an elevated stone platform. The two of us marveled at its fat belly, and the entire image, its deep pedestal sunk into the platform masonry, was some ten feet tall, carved entirely from one block of red stone, a kind of stone that did not exist anywhere locally. Aletti told us that the quarries were far up the valley, where some unfinished blocks of this very same stuff had been discovered, together with the hard basalt adzes used in carving them.

The red statues had obviously been raised on some outdoor temple ground. We began looking around the underbrush and found that there were walls and terraces everywhere. A few statues lay half buried in the ground, their heads or arms broken by force. There were also some huge round heads lying on the ground, gazing at us from between creepers and ferns. These heads had been carved as independent stone images without torsos or bodies. The strangest of all the sculptures was a bulky figure carved as if in a swimming position, with stunted arms and legs stretched out fore and aft, the whole resting on a short pedestal extending from the abdomen to the ground as part of the same piece. It was carved from a hard, fine-grained rock and represented the absolute perfection of design and workmanship in stone. Only the best professional stone sculptor could create a work so masterly and so symmetrical, streamlined and polished.

Never before had I seen a monolith carved to represent somebody lying on their belly with arms and legs stretched out in opposite directions. It was utterly un-Polynesian, and nothing remotely like it had been carved on any other of the thousands of islands in the entire Pacific.

To check every detail, I tore away the grass and turf around the supporting column to see how deeply it continued into the ground. Aletti helped me with his big knife. To our astonishment, we discovered reliefs all around the base of the short column. Two squatting human figures with hands above their heads appeared fore and aft, and on either side a mammal with a long body and thin, erect tail. The heads of these animals had blunt muzzles and round erect ears, and the eyes and mouths were clearly incised.

Two quadrupeds! Here was something for a detective. Every student of Polynesia knew that there had been no mammals on any of the Marquesas Islands in pre-European times except rats and pigs. No rat had a neck like this, or an erect tail. And this was certainly no pig, for a pig would have had a tiny piece of a tail curled up like a stump of rope at its rump. There were dogs on most of the other Polynesian islands when the Europeans came, so the dog had probably been brought to the Marquesas also and later been lost. But the Polynesian dog, *Canis maori*, had a thick, bushy tail curved in a permanent circle, not standing straight up like a stick as it does on a cat. A cat? Yes, it could depict a cat, but there were no felines, neither wild nor tame, in any part of Oceania, Australia included.

The nearest feline was the puma, the sacred symbol of divinity and royalty in ancient Peru. The puma was common in pre-Inca religious sculpture, and two pumas with raised tails were carved in relief on either side of the monolith of the pre-Inca sun god in Tiahuanaco. Certainly this could represent two felines. But that would indicate some contact with South America. And all authorities agreed that South America was too far away from Polynesia for any contact to have taken place before the Europeans came.

And yet. South America was far away indeed, but that was where wind and current came from. And it was by far the nearest mainland.

Anyhow, with little effort I had made a new discovery, and even the local islanders came up from the coast to look at something they had never noticed before. I learned from them that they had actually reerected this stone image themselves only a few years ago, as it had been laying capsized in the bushes, probably overturned by the Hawaiian missionary Kekela. He had converted their ancestors to Christianity fifty years ago, and planted coffee shrubs at the temple site, which had been used for cannibal festivities until then. In fact, most of the underbrush that had overgrown the area surrounding the image was still sprinkled with red coffee berries.

As the curious natives came up the valley with Henry Lie and the little Frenchman, and they were made aware of the animal reliefs, they all had to revise their former belief that this swimming statue represented a woman giving birth. According to Henry Lie, until quite recently women had secretly brought food offerings to place before this image when they were expecting a child. He could also tell me that three scientists had visited this temple site. F. W. Christian and Karl von den Steinen in the 1890s, and more recently the American ethnologist Ralph Linton in 1920, when Henry Lie had taken him to see the images. I knew the books all three had written from my studies in the Kroepelien library. They had all been told different tales and been given different names for the statues by the local natives, who admitted to Henry Lie that factually they knew nothing. I was the first visitor to see this image in its right position. All the others had seen it upside down. The short column under the belly of the image, serving as its supporting pillar, had then projected meaninglessly into the air and was assumed to be the coarse representation of a child emerging from a woman's womb. Nobody had noticed that the child emerged from the navel area rather than from between the legs.

To Linton, the last of the three scientists, the people in the valley had actually confessed their ignorance. They told him what he also put on record: the statues were already there when their own ancestors had arrived on the island and driven an earlier people away into the mountains. Who these earlier people were, nobody could tell Linton, although certain traditions maintained that the old Naki tribe had absorbed the blood of these earlier stone sculptors.

Since Linton was told by the island population that their ancestors had not carved the statues, it was difficult for him to believe that they knew what the images actually represented. In his book *Archaeology of the Marquesas Islands*, he clainied that the image with the outstretched arms and feet differed so much from all the other statues at the site that it could scarcely be identified as human. He had no idea of what it could represent, but wrote: 'It is evident that the artist worked on a preconceived plan and was a master of design and execution.'

There are incidents in everyone's life that may be casual and yet prove to have vital consequences in future development, even to the extent of redirecting an entire life. My introduction to the Puamau stone giants during an attempt to return to nature happened to switch me onto a new track that was to guide my destiny for many eventful years to come. It set me assail on rafts, led me into the jungles of several continents, and made me excavate the hidden body of Easter Island stone heads that towered as high as four-story buildings and made the Marquesan stone giants seem like dwarfs.

There was an immediate sequel to the confrontation with the mute stone figures. The two autodidacts down by the beach helped the stones to talk. The same evening Henry Lie placed a stack of well-read books next to the bright kerosene lamp, and began to turn pages I had read before but perhaps with too much confidence in the assumed authorities. I had already seen things at sea and in the island jungles that had begun to shake my confidence in some of their generally accepted conclusions. For months I had felt the westward-moving elements of the Pacific on my bare skin, and I knew that to the Polynesians west was 'down' and east 'up.' And yet the authorities agreed that all the original settlers of Polynesia had come from the west, none from the east. None from the coastal civilizations of South America. Why? One thing was abundantly clear: the mystery of the origins of the Polynesians was by no means solved. No two of the authors who had dealt with the question agreed. No two of the many dozen who had advanced a theory had come to the same conclusion. They all disagreed and disproved each other. Only one thing did they all seem to agree upon – the Polynesians were of mixed racial and cultural origins. More than one group of voyagers had reached the Polynesian island area, at different times and along different routes. That conclusion was based on a variety of observations and was supported by the historical traditions and genealogies of every part of Polynesia.

'There is no reason to doubt what these people were absolutely sure of,' said Henry Lie. 'Two different peoples had come to this island before any Europeans. The Polynesians were ancestor worshippers. If their own ancestors had erected these monuments they would never have forgotten it. On this island all important families kept genealogies by means of a system of knots on string bundles called *kipona* and resembling the quipu of Peru.'

He reminded me that throughout Polynesia there were persistent traditions to the effect that another people had been living on the islands when the ancestors of the present population arrived. Everywhere within the Polynesian triangle, which extends from Easter Island in the east to Samoa and New Zealand in the west and to Hawaii in the north, the learned men of the tribes agreed: an industrious people, often with reddish hair and fair skin, had been found by their forefathers on the various islands and expelled or absorbed by the newcomers. To the Europeans these claims were fairy tales. To the Polynesians this was reality. The traditional memory was still so vivid that the first Europeans were mistaken for returning groups of these early light-skinned people. Captain Cook lost his life in Hawaii when the mistake was discovered. He was not as lucky as Cortez and Pizarro, who conquered the vast Aztec and Inca empires in Mexico and Peru without a fight, because of the same strange belief in an earlier visit by white-skinned culture-bringers.

And wherever in America these legendary fair people were remembered, *they had left behind large stone statues*. All the way from the Gulf of Mexico down through Central America, and then on down the pacific side to Colombia, Equador, and Peru as far as to Tiahuanaco. They appear again in the open Pacific, but only on the easternmost outposts as Easter Island and the Marquesas, the very island directly facing South America.

Then Henry Lie surprised me: 'The ancestors of the present Polynesians must have come from forested coasts,' he argued, 'for they were great wood-carvers. They carved totem poles, split planks, and carved the bows and sterns of their canoes. Nobody ever saw a Polynesian carve rock.'

The Frenchman agreed, and eagerly poked tobacco into his nostrils as he pulled forth a large volume. He found a page with a picture and pointed. 'Look at this.'

I looked, and was amazed. He was showing me a picture of a stone statue amazingly like those standing up in the valley. I looked at the cover. It was a book about travels in South America.

The picture was of one of the great number of giant stone statues scattered about over wide areas in the jungles of San Agustin, in Colombia, the nearest land due east of the Marquesas. All the features of this South American stone giant were the same as those characteristic of the images we had seen that morning: the colossal head making up one third of the total height, the ridiculously stunted legs making up another third, the hands clasping the fat belly, the intentionally grotesque face with enormous circular eyes, flat, broad nose, and an extremely wide mouth with thick lips grinning from ear to ear.

'Look at the arms,' said the Frenchman enthusiastically. 'Every single statue on this island has the same peculiar pose, the elbows bent at right angles and the hands held on the stomach.'

I looked at the pictures of a few other stone figures from San Agustin, then went outside to feel the breeze of the trade wind. It came full force to this remote valley with its lonely stone statues from the continental area where their counterparts were rooted in Andean culture. No wonder the Polynesians ascribed them to an earlier people. But South America seemed too far away. Yet in the opposite direction there were no stone statues of any kind until those of Asia, twice as far away and against wind and current. The two nonschooled settlers and the mute stone men in their valley had made me sceptical of the one-track mind and indoor map-reading of most Polynesian scholars.

The old Frenchman snapped the book shut triumphantly, as if closing a jewelry box after letting me admire the treasure. He could see that it had made a deep impression. I did not know what to believe. All the textbooks held that a sea voyage from South America to these islands was impossible until the Europeans introduced plank-built ships. That was just about the only thing everybody agreed upon. But not these two unlearned characters, who were sitting here pondering for themselves, far from any scholarly society. They looked toward South America.

I spent a whole week up at the Puamau temple terraces, known as Oipona. No nook or cranny in the area was to be left unchecked. But it was not on the island itself that I was to find further sculptures by the artists who had carved the Oipona images. When the German explorer von den Steinen had visited the site in 1896 he had removed the finest and best preserved of all the Oipona stone heads to the Völkerkunde Museum in Berlin. I had been to that museum with my father as a student, while I was preparing my research on the Marquesas group. I had seen the stone head then, but at that time I had not had any reason to study it carefully. But years later, when I had personally seen the whole assembly of stone giants in the Puamau valley, I came back to Berlin to take a better look. Now I detected that the big head in the German museum had relief decorations all around its neck, and the motif was precisely the same I had discovered around the abdominal cylinder under the prone image of Puamau. The same two squatting human figures alternated with the same two long-tailed quadrupeds. But on the head von den Steinen had brought to Berlin, the animal figures were even better preserved, so well indeed that one could see the long claws on the feet and the whiskers at the snout. The attempt at depicting a feline and not a pig, dog, or mouse was now quite obvious. As felines were not part of the Pacific Island fauna, and the Polynesian dogs and pigs did not have long straight tails, von den Steinen had concluded that the animals, in spite of their fierce claws and erect tail, had to represent rats.

By the second time I saw the feline claws and whiskers on the twin animal reliefs on the Marquesan head in Berlin, I had become familiar with the monumental art of stone-carving seafarers in both the Old and the New World. Nobody had decorated sculptures of divinities or deities with rats. But felines indeed. Felines, commonly in pairs, appeared as an accompanying motif on monuments of divine rulers among some of the earliest stone-carving civilizations in both the Middle East and South America. Among the Hittites, a pair of lions were commonly sculpted as separate monuments flanking the sides of a divine image. In the Andes, a pair of mountain lions, or pumas, took the place of the Old World lion in their religious art. In Tiahuanaco, a puma is carved in relief on either side of the seafaring sun-god Kon-Tiki Viracocha.

But the swimming image of Puamau, with the two felines decorating its base, had a remarkably restricted distribution. I had to go to the jungle site of San Agustin to find a match. There, among the innumerable standing images of diabolic men, were two in a horizontal position, grinning at me. Each had a huge mouth and large eyes, and their bodies were stretched as if swimming, their stunted arms stretched forward on either side of their faces. The concept was clearly the same as on the prone creature in the Marquesas. These continental counterparts represented the swimming cayman god. Neither caymans nor felines belonged to the fauna of the Marquesas. But somehow their concepts had found their way over the sea to Hivaoa before the present people came.

*

Twenty years of Polynesian research would pass – with excavations in the field and studies in libraries and museums – before I came back and met Henry Lie again. Anchoring off Puamau in my own expeditions ship, I was to learn from the natives ashore that a terrific flood had swept their valley. Torrents of water had carried the bungalow of Henry Lie and the hut of the Frenchman into the ocean. Gone were Lie's books, and gone was his fine collection of ancient images and other relics. The Frenchman was no longer there. But Lie had survived and moved to the next valley, where we found him busy clearing the overgrown wilderness for a new plantation.

Only the bulky red stone giants up in the valley stood there immobile and unchanged, once again partly hidden by jungle foliage and coffee bushes. How long had they been standing like this?

That was what I had come back to find out. Nobody knew how old they were. And nobody knew as yet the age of the still larger stone giants on Easter Island. But all authorities took it for granted that the idea and skill in carving and erecting stone giants had first evolved in the jungle islands of the Marquesas, and then spread from there to lonely Easter Island. This assumption was based on the conviction that Easter Island must have been the last island to be settled, since it was the one Pacific island that lay farthest from Asia. If it were certain that all voyages into the vast Pacific hemisphere had come out of Asia, this reasoning would make sense. But nobody had thought of the possibility that somebody could have come the other way.

On my return to the Puamau valley, I brought along a team of professional archaeologists. We came from Easter Island, nearest to South America. We anchored off the coast with our own ship, which we had chartered for a year, and had already collected ample material for radiocarbon dating of the Easter Island giants. Now we wanted to look under the platforms at the Oipona temple site to determine when the Marquesan monuments were erected.

Back in Europe a year later, the carbon samples from both areas revealed that the Easter Island stone giants antedated their smaller relatives in the Marquesas by at least half a millennium. The Puamau images were erected about A.D. 1300, whereas the first colossal stone men had been set up on Easter Island by A.D. 800.

Thus, if there had been a direct relationship, the inspiration must have spread from Easter Island to the Marquesas rather than the other way.

But *was* there a direct relationship? Perhaps. But I was not at all sure. We had dug up the oldest types of statues so far known from Easter Island. They were incredibly similar to the oldest types from Tiahuanaco. The statues of Hivaoa were very different. They resembled the prehistoric monuments of San Agustin. The difference in latitude between Hivaoa, which is close to the equator, and Easter Island, which is much further south, is the same as the distance between San Agustin and Tiahuanaco respectively in the northern and the southern extremities of the Andes. Could it be that stone sculptors from these two widely separated areas of the Andes had sailed west with the trade winds independently and influenced the two separate outposts of Polynesia closest to their own continental domain?

None of this was known to Liv and me as we saddled our horses to ride with Terai back over the mountain crests to Atuona Valley. I had not yet visited the sites in the Andes nor dug up the early types of statues on Easter Island. No archaeologist had as yet been digging in any part of Eastern Polynesia, and I had no idea that I should become the first to bring them in.

From Puamau I carried a secret dream of challenging the dogma that seemed to have blocked all unbiased research in Polynesia: the unproven axiom that only aboriginal voyagers from

Asia sailing across the ocean could have reached Polynesia.

All along the Pacific coast of the Inca empire, from Colombia, Ecuador, and Peru to northern Chile, there had been numerous as yet little-known pre-Incan cultures basing their economies on coastal navigation and fisheries. Why should they be totally ignored? To Polynesian investigators, the Pacific seemed to have only a western shoreline. To the east there was an impenetrable abyss. It was as if America had not yet been placed on the map by Columbus.

XI

The Island of Environmental Holocaust

WE DECIDED TO RETURN TO FATU-HIVA. On the way back we happened to make a call at a little island that I will never forget. An island of ill omen that forever left an impression of horror in my memory. And yet this was the very island that had become the entrance to the Polynesian paradise when the first Europeans discovered that there were inhabited islands in the vast expanses of the Pacific Ocean.

Two weeks had passed since we rode back along the mountain crests from the valley of the stone men before we got a chance to return to Fatu-Hiva. Two weeks in the verdant valley of Atuona that inspired Gauguin and enriches all those who treasure his art. To those who lived there now, the valley's value was estimated in copra sacks. There was something we could not digest in the pseudocivilization that had taken possession of this marvelous environment. The colors still existed, but gone were the eyes that saw them. Atuona was the administration center for the southern islands in the Marquesas group, and was visited by a French cruiser from Tahiti while we were there. The governor of French Oceania stepped ashore for a moment, accompanied by officers all in spotless whites. They were received by a double row of hula dancers, also all in spotless white, but with green grass skirts dangling over the men's trousers and the women's gowns.

When the cruiser was gone, we found that somebody had left a wreath at the modest tombstone of Paul Gauguin. Somebody who appreciated what this environment had meant to him.

We could have left for Tahiti with the cruiser, but Liv refused any thought of asking for a lift. It would be stupid to give up Fatu-Hiva without another try. As we watched the gray sea monster move away, guns fore and aft, we were as certain as ever that our planet was ripe for another war and we two would get involved if we returned to civilization. We did not want to be eye-witnesses to modern society's collapse under the burden of all its tanks, guns, and battleships. Armed with nothing but the butt end of Gauguin's gun, we would confront no enemies on Fatu-Hiva. None but the mosquitoes, and it was better to be bitten than bombed.

The battleship had disappeared by the time the trading schooner *Moana* anchored beside the promontory of Atuona Bay. The captain had come for copra and was willing to take us back to Fatu-Hiva.

We were the first passengers to come on board, and installed ourselves on the hatch with our sparse belongings. The loading of copra sacks continued until dusk, but passengers still arrived after we had fallen asleep.

I woke up to the rattle of heavy anchor chains, and as I opened my eyes I looked straight into the face of a beautiful Polynesian girl lying by my side. I sat up and saw Liv lying sound asleep on my other side. I had to lie down again in a hurry so as to avoid a blow on the head from a big boom that swung over my face. The unknown beauty laughed, her black hair blowing all over her face, and wrapped herself up to the nose in a blanket. At my feet, the early morning sun swung slowly to the side as the schooner turned, then disappeared behind the sail as the boom passed over us. Pitching turned into heavy rolling, and we were on our way out of the sheltered bay.

With the large wind-filled sail forcing the boom steadily to starboard, we could all sit up and enjoy the sight of the *Moana* gliding like a white eagle out of the green landscape, before Atuona Valley dwindled and disappeared. Farewell, land of Terai, of Gauguin's tombstone and the standing and swimming stone men. Little did I know that I would come back one day with my own expedition.

As we rounded the point of the promontory, we saw the ruins of what had been the former leper station. Leprosy had been the most terrible gift white man had brought to these islands. The buildings had recently been set on fire and leveled to the ground, in order to kill the bacilli and arrest contagion. The remaining inmates, we were told, were sent home to their huts, except some who were interned in the leprosarium on Tahiti.

The never-failing trade wind hit us full force now, and we sniffed in salt-fresh air coming from across the ocean. Some frozen islanders withdrew below deck and the rest of us wrapped the blankets more closely around us. Three young girls on my left were itching to make acquaintance, wriggling and giggling in Polynesian fashion. But they became calm and serious when those who had gone below deck, came up again with news that spread from mouth to mouth: there was a madman with us. Belowdecks was a dangerous prisoner on his way to jail in Tahiti. He was cursed because he had broken the taboo.

The captain had the details. Two men had come from Tahiti to try to resettle an abandoned valley next to Atuona. In the valley there was a great old tree known to be taboo. The

Tahitians were not frightened and climbed into the tree. It was hollow. Inside were three human skulls, the largest they had ever seen. They wanted to sell these skulls in Tahiti for a high price, and ignored the warnings about the taboo. They hid the skulls in a suitcase for shipment with the *Moana*, but at night the heads began to whimper inside the suitcase. One of the two repented and wanted to return the skulls to the tree, but the other refused as he wanted the money. Then the noise from the suitcase became so loud that it woke up everybody in the house. The fellow who had refused to replace the skulls in the tree became so frightened that he went out of his mind. He drew his long machete from its sheath and ran in pursuit of his friend to cut off his head. Others interfered and a mad fight followed before the lunatic was disarmed and dragged by force across the mountains to the *gendarmerie* in Atuona.

Now that he was below the hatch we were sleeping on, I could not help thinking of the little fruit-rat that had entered one of the waggling skulls below our bed on Fatu-Hiva. Perhaps these innocent little Marquesan rodents had a habit of nesting inside the empty braincases of old human skulls?

Everybody sat up as someone noticed that the island of Motane had risen above the horizon on the port side of *Moana*'s bouncing bow. We recognized the hazy contours of the same island we had passed at a distance on our way north in the lifeboat. Then we had passed with no thought but to reach Hivaoa alive. This island was known to be without any population. We were therefore surprised when the Tahitian captain of *Moana* ordered the helms-man to lay the wheel over and bear straight down upon an uninhabited coast.

We needed provisions, he said. Fresh meat.

I was pleased to get this unexpected chance to visit Motane. Historically speaking, this was not merely one of the tens of thousands of islands in the Pacific hemisphere. This was where the European history in Polynesia began. Fatu-Hiva and Motane were the first two islands in Polynesia to be sighted and visited by the Spaniards dispatched by the viceroy of Peru to search for inhabited islands known to the Incas. Two centuries before Captain Cook's important explorations in Polynesia, the learned *amautas* of the Inca empire had told the Spanish conquistadores about people living on islands far out in the Pacific Ocean.

On 19 November, 1567 two Spanish caravels left Callao harbor in Peru with an expedition of 150 men who had been ordered to visit and convert to Christianity the Pacific islanders known to the Incas but not yet to any Europeans. The viceroy's nephew, Alvaro de Mendaña, was appointed commander of the two ships, and the party included the famous Inca chronicler Pedro Sarmiento de Gamboa, who had obtained exact bearings from the Incas and taken the initiative for the enterprise. The chief navigator was Gallego, a longtime pilot along the coast of the former Inca empire, with independent information from the contemporary balsa raft voyagers and the *Cancilleria* in Lima. Sarmiento's sailing directions, based on information from the Incas, was 600 leagues (slightly more than 2,000 nautical miles) west-southwest from Callao, which would have brought the two ships almost exactly to the location of Easter Island.

But after ten days, when the ships had barely passed latitude 15 degrees south, quarrels began between the Inca historian and the chief pilot, and the course to the island supposed to lie still further to the west-southwest was interrupted. Gallego had been told that most of the inhabited islands lay about latitude 15 degrees south, and as the expedition commander put more faith in the pilot than in the Inca historian, the ships altered course, to Sarmiento's despair. If Gallego had been consistent, and faithfully followed his own informants' advice to steer westward along 15 degrees south, the Peruvian ships would have headed straight into the heart of Central Polynesia, and landed in the Tuamoto archipelago or the Society Islands, which included Tahiti. The concentrated belt of Polynesian islands is, in fact, clustered between 10 and 20 degrees south. But the fitful pilot altered course a second time, and passed midway through the open gap between the Tuamoto archipelago and the Marquesas. The expedition thus bypassed Polynesia and finally landed among black-skinned Melanesians in the distant Solomon group. So – to the outside world – in 1568 these post-Columbian voyagers from Peru became the first discoverers of Melanesia. That pre-Columbian voyagers from Peru had visited Melanesia before them, however, seems highly likely. For the Inca historians and sailors, who gave the Spanish conquistadores the direct sailing directions to inhabited Pacific islands, had predicted that they would find islanders with black skin. Firm Incan traditions, carefully recorded by Sarmiento, stated that Inca Tupac Yupanqui, the grandfather of the living Inca, had sailed for nine months in the open Pacific with a large fleet of balsa rafts to visit islands known to the coastal population of Peru. He had found some islands with a black-skinned population and brought some of them back as curiosities to prove that he had personally reached those distant shores. The arriving Spaniard recorded that they had even been shown some of the souvenirs the great Inca had brought back from the islands.

Fatu-Hiva and Motane came next. Thanks to the fitful maneuvers of the chief pilot on the first Mendaña expedition, Polynesia remained unknown to Europeans for another twenty-six years. Then a second Mendaña expedition set out, again from Peru, and ran into Polynesia in 1595, when they landed on Fatu-Hiva and Motane. Here the Europeans met Polynesians for the first time and described them as fair-skinned and handsome, some even with naturally red hair. One of the Fatu-Hiva women was described as having such long and beautiful red hair that the captain was tempted to cut off a lock, but refrained for fear of offending the islanders.

History repeats itself, I thought, and the European voyages from Peru were in mind when we headed for the leeward side of Motane to find shelter from the eternal trade winds from South America. Our own medieval history had a lesson to tell us. Although European traders and settlers had reached the Pacific shores of Asia two centuries before any Europeans came to Peru, not a single island in the open Pacific was encountered before Columbus had led the way to America. Then all of Oceania immediately lay open to the sailing ships. Even the oceanic islands right next to the Asiatic continent were discovered from America. Before Peru was reached by Pizarro and the first Spaniards, Magellan had rounded the tip of South America and crossed the entire Pacific with favorable winds and currents, discovering Micronesia in 1521, when he landed in Guam, the very island nearest the old European settlements in Asia.

We were now approaching Motane, which rose out of the mist and took on three dimensions. I had read what Pedro de Quiros, chief pilot on the second Mendaña expedition, had written about Motane: 'It is an island beautiful to look at, with much woods and fair fields.'

I expected to see the hazy contours transform into a verdant green wilderness, and was amazed at what seemed to come gliding toward us as a white ghost island. Gone were the woods and the fields.

Moana anchored, and with the lifeboat we ventured in on the leeward side as close to land as the restless surf permitted, then jumped ashore onto some rocks at the foot of low cliffs. Ashore with us came the Tahitian cook with his gun and half a dozen other Polynesians, crew and passengers, while the others took the lifeboat safely back to the schooner, waiting for us to signal for our return.

Stripped but for colorful pareus knotted like long G-strings between their legs, some of our Polynesians dived into the surging water with slender spears, looking for spiny lobsters and fish. The rest of us climbed the coastal cliff and proceeded up the hot, sun-baked slope where the sharp light blinded us. The ground was as white and dry as dust, covered with sterile sandy soil, loose stones, and bare rocks. Scattered far apart were clusters of dry shrubs with thick, sapless leaves. The whole island was otherwise devoid of a green blade or tree. The sun scorched the dry ground, unhindered by any crowns of forest, which was all gone. The landscape had been transformed into a desert. At considerable distances from each other, we saw some dead trees, as white and dry as bone, sapless and stripped of both leaves and bark, their naked branches stretched like claws toward the blue sky.

Scattered everywhere were bleached bones and complete skeletons: twisted horns of rams, animal craniums, ribs, and leg bones. Everywhere we moved in the rolling terrain, we saw the same sight: windswept stones, dry scrub, and distorted skeletons of sheep. In a dead tree inland a wild cock sat crowing, and another answered in the far distance.

Man had once lived here, as we knew. In several places we came across nicely built *paepae*, old Polynesian house platforms. In the Marquesas, the houses had to be raised above the jungle mud. There was no mud here now, but in the gullies and glens we saw vestiges of watercourses and potholes from former streams and rivulets. There was not a drop of water now, not even the dark shadow of humidity. The salty surf against the windward side of the island seemed to augment our longing for drinking water in this hot, arid landscape.

We needed no guide to find our way around in this open terrain, nor did we need anyone to tell us what had happened. Here in Polynesia the white man had discovered what he described in early logs and diaries as a happy and carefree people living in harmony with a healthy and lush environment. But the white man was cash-oriented and tried to improve nature with copra plantations and imported livestock. In his wake came disease, and when white man saw his own shadows he withdrew. When Captain Cook ran into the Marquesas group in 1773, he

I met Liv at a graduation party, and found that she loved nature as much as I did. While still university students, we made our plans to escape civilization and return to nature.

Happy days in Polynesia, where both big and small are children of the sun.

Signs of true civilization had begun to appear on the coral atolls of the Tuamotu group. There were still no roads throughout the archipelago, but the first automobile—a prestige symbol—had already arrived.

Liv and I were set ashore on the empty beach of Omoa Valley on Fatu-Hiva, left alone in a totally foreign world as the schooner sailed for distant Tahiti.

Before the sun set and left us in the jungle darkness, we had cleared enough space to put up our little tent.

We built our first home, of plaited bamboo and coconut-leaf thatch, on top of the stone platform that had once held the home of the valley's queen. Our kitchen had a roof set on poles.

The river that ran down the bottom of the valley was only a short distance from our bamboo cabin. It was here we enjoyed our happiest and most relaxed moments.

On rare occasions, we were visited by islanders from the villages down by the coast; some of them showed us where human skulls and artifacts had been hidden by their ancestors.

I had to fetch our daily bread from the jungle as there was nothing we could buy for money. I often found, in the places where I discovered bananas growing, the ruins of stone platforms on which houses had once been built.

I felt like a king when an islander brought me an ancient royal crown; it was made of Tiki images in turtle and conch shell, and tied to a band of coconut fiber. Liv, also regal, took over the former queen's beautiful bathing place.

Along with a variety of archaeological artifacts we collected a few ancient skulls. We became somewhat horrified one night when they all began to rattle under our bed: a fruit rat had crawled inside one of the craniums.

Large mango trees planted long ago by the islanders grew at intervals along the trails in the open highlands. They gave us and our horses a welcome rest, providing shade and deliciously ripe fruit.

The air was always fresh and cool in the highlands; all the mosquitoes were blown away by the constant trade winds. We felt reborn each time we left the damp jungle valleys to ride on the roof of the world.

Our unshod horses were caught as foals in the mountains and domesticated. They were tireless and surefooted on long and strenuous trips over the narrow cliff.

When we returned to our abandoned home in the jungle of Omoa after a month's absence on Hivaoa, we found that the poles of our kitchen shelter had started sending out long, leafy branches from their tops, and roots had crawled into the fertile mold. The bamboo cabin had collapsed and begun to turn into white dust and black humus, ready for transformation into new life.

Old Tei Tetua was the last and only survivor of all the tribes that had inhabited the windward side of Fatu-Hiva in pre-European times. He had become a Christian and had dug his own tomb, into which he planned to climb just before his death as a precaution against being eaten by wild dogs.

We built our new home down by the beach, next to old Tei's cluster of cabin
It resembled a bird's nest—a palm-thatched roof over a bamboo floor th
was raised so high on poles that we reached it by a ladder.

Tahia-Momo was Tei's adopted little daughter, brought over from the other side of the island by her family to keep the old man company when he became lonely. Tahia and Liv, despite a few years' age difference, became real friends.

The splashing surf created countless pools in the strange lava formations on either side of the bay. Here Liv and Moma found a colorful and incredibly varied collection of shells, fish, and sea anemones. There was no danger of starvation in Ouia Bay.

Like his ancestors, Tei Tetua lived in harmony with the nature that fed him. The stone blade of his adze had been exchanged for an iron one, and he acquired a flint with which to strike sparks for fires instead of rubbing two sticks together. He cultivated nothing, yet fed us and himself generously with food from our undomesticated environment. Tei entertained us by playing his bamboo flute—with his nose.

We caught a wild goat and brought it to our cabin, hoping for milk. But it was a billygoat.

One day an Omoan goat-hunter came over the mountains with his pack of dogs. It proved to be our friend Veo, who wanted to find out what had happened to us. He was a truly fine Polynesian of the old type, and we as well as Tei welcomed him and hoped he would stay in our marvelous valley.

Since Veo never returned to Omoa with news of our disappearance, more and more people crossed the mountains to enjoy Tei Tetua's hospitality, making him feel that his world had been reborn in all its past happiness. But among the newcomers were real troublemakers.

When Liv and I woke one night to find that a giant poisonous centipede had been put in our bed, we decided to escape before daylight. We climbed the steep mountain crest as the tropical sun rose to its zenith, and we were almost roasted alive along with the tiny piglet Liv had carried with her to keep as a pet.

Since my first arrival on Fatu-Hiva, I had learnt that the local fishermen feared the snakelike teeth of the moray eel more than the sawlike teeth of the shark. Little did I know then that we should end our sojourn on Fatu-Hiva among the boulders of a cave on the beach, where moray eels had to be chased away from our bed.

Some important cultivated plants that could not have moved across the ocean by themselves must have been brought from South America to Polynesia in pre-European times.

There were no boats in pre-European times in South America—only the balsa-wood raft. Therefore a balsa raft had to be able to sail from South America to Polynesia.

estimated a local Polynesian population of one hundred thousand, but after him came European settlers, whalers, merchants, and missionaries. In 1883 the total census for the group had shriveled to 4,865. When we had arrived on Fatu-Hiva, we found dozens of formerly inhabited valleys abandoned. The great majority of the valleys in the Marquesas group are, in fact, abandoned. The wilderness has overtaken house foundations and former fields.

Not so in Motane. Man and environment had both perished together.

As we advanced to the higher terrain in the center of the island, we saw the first sign of life on the ground. A few frightened sheep ran bleating with their lambs through some dry bushes. They were scraggy and undersized, with scanty wool. The three crew members in our company ran barefoot after them uphill and downhill, and the clearly feeble animals could not escape. The gun was not needed. The unarmed pursuers just threw themselves upon the sheep, grabbing them by the wool. Soon Liv and I were left alone, meditating on the tragic sight, while the three men marched back toward our landing place, each carrying a bleating sheep like a bulky stole around his neck. Laughing victoriously, the one forming the rearguard of the party turned around and told us to wait; they were coming back for more. Evidently there was little flesh under the skin of their booty.

Never before had the sun, the very intensity of the sunlight, given me the same feeling as when a full moon shines on a cemetery. The white trees stood like tombstones over a pillaged graveyard; there were skulls and bones everywhere. It was midnight at noon.

We sat down on the smooth stone platform of an old *paepae*. The walls of a family home had once stood on top and provided shelter and shade from the scorching sun. There had been no windows but a low door in the wall, where children once ran in and out, where men stooped to pass in with fish on their spears, and the scent of poi and baked breadfruit seeped through the plaited palm-leaf walls. Perhaps the naked ghost tree before us that gave us no shade had once been heavily laden with fruit.

This was indeed the stunning island with woods and green fields described by the first Europeans. With a little imagination we could see our own schooner transformed into Mendaña's caravel, anchored down in the blue water partly hidden by trees. Here men with three-cornered hats and tight-fitting breeches had once come ashore with blazing firearms to look for water, fruit, and women. They brought rum and the recipe for brewing drinks far more potent than the innocent kava drunk by the islanders. They brought a different living system, refined sugar, white flour, and disease. To a people who knew no other domesticated mammals than their own hairy pigs, the visitors brought goats, which the islanders believed were pigs with tusks on their heads. The islanders accepted all with gratitude and admiration and offered their fruit, poultry, and women in return. A new religion was accepted, too, and the era of Polynesian savagery ended. The Polynesian era ended. On Motane, human history ended.

Perhaps it was in the cool hut that once stood on the sun-scorched *paepae* where we were now sitting that the last inhabitant of Motane had closed his eyes. Or perhaps the last survivors escaped by canoe to Hivaoa, whose mountain ridges could be seen in clear weather. Maybe the final survivor was a child, left alone when the imported goats and sheep ran wild into the green wilderness.

To me, Motane was a terrifying example of what would happen if nature was tilted out of balance. The Polynesian environment was clearly not ideal for the birth control of sheep. On a continent, carnivorous beasts would have helped reinforce nature's law of equilibrium. In a balanced environment, the composition of species is precisely calculated to permit each kind of animal to have two of its offspring grow to maturity for each pair of parents. No matter if the descendants come as lambs, tadpoles, bird's eggs or fish roe. If less than two survived, the species would gradually die out. If more than two survived, the forests or the sea would be overpopulated by that one species. An automatic birth control is one of the wonders of nature. There are thousands of eggs in the spawn of a single salmon, but only two succeed in becoming adult fish; if this did not happen, there would soon be no room to swim.

But when the last of the flesh-eating humans left Motane, there was nothing to stop the plant-eating sheep from breeding lambs and multiplying far beyond the saturation point. Hordes of wild sheep consumed all the grass, all the leaves within reach, and when famine hit them, they devoured the roots of the grass and the bark of the trees until even the last of the foliage high above their heads withered away and the island became a desert. Without trees to shelter the soil from the scorching sun, without roots to hold humidity near the surface, every drop of rain sank deep into the thirsty ground and was lost long before it reached any glen or watercourse. The gushing streams lost all their supplies, and the last rivulet disappeared from the surface of the land. Motane's biological clock had not only stopped, but was set to go backward until the hands showed a visitor pretty much what our planet looked like before life emerged from the sea. If the *Moana*'s crew managed to catch the last of the scraggy sheep, then the miniature world we saw from our hilltop would more or less match Planet Earth in its early formative period: that remote era when the ocean and the air were filled with creatures swimming and flying around the otherwise lifeless coasts. But if all the world were reduced like Motane, how many million years would we have to sit alone on our hilltop *paepae* waiting for evolution, for algae to be washed ashore and develop into grass and trees once more, or for fish to jump ashore and acquire lungs and legs and fur a second time? The tiny fish we had seen jumping in the surf area on the cliffs of Fatu-Hiva had certainly jumped like that for hundreds of thousands of years, and not one of them was as yet ready to take the first leap on the long road toward kangaroo or monkey. Better to take care of the world we had. It would take a long time to get a new one.

We were almost asleep on the old *paepae* when the Tahitian crew gave up the hunt, and we were all ready to return to the *Moana*. The clear waters around Motane were evidently still teeming with marine life. The marvelous variety of seafood speared by our divers included several fat moray eels, which, cooked in the milk of coconuts brought from Hivaoa, provided passengers and crew with a meal superior to that cooked from the skin and bones of the starved Motane sheep.

Before the sun set, we left the dead island behind us. The brief visit seemed a long voyage through time. Backward, to an incomplete planet in the making. Or forward, to the horrors of a planet where man's mismanagement of his environment had turned green leaves and red blood back to the gray dust from which the miracles of evolution started.

Two thousand miles to the southeast of Motane and Fato-Hiva, modern technology would one day give me a new chance to travel back through time. Since early boyhood I had been fascinated by the reports of the colossal stone heads that emerged from the soil of Easter Island, the loneliest speck of inhabited land on our planet, half way between South America and the nearest outposts of Polynesia. The subtropical climate of this island should have granted its rolling landscape and extinct volcanoes the same lush vegetation as the other South Seas islands. But neither palms nor forest trees, only rows of lofty stone giants rose above the ground when Admiral Roggeveen reached the island from Chile on Easter Day, 1722. This mysterious island was then as barren as the windswept landscape of the islands north of Scotland.

When my converted Greenland trawler anchored off Easter Island's east coast with a team of scientists in 1955, we witnessed the same barren landscape as seen by all visitors before us. Not a single tree could be detected ashore.

Had Easter Island looked different before the sculptors of the giant stone men came ashore? Where had they come from? Who did it? When? How? Why? One of the problems we had come with means to solve was whether the pre-European stone sculptors had destroyed a former forest. How could all the incredible engineering accomplishments have been possible without timber?

So, we came to this island with tools enabling us to penetrate the distant past. Even permitting us to get a very good view of how Easter Island's vegetation had changed through the ages. We brought a twenty-foot long drill to attempt pollen borings around the Easter Island crater lakes. A pioneer in Pacific Island palaeobotany, Dr. O. H. Selling of the National Museum in Stockholm, had provided the expedition with all we needed to bring him plant pollen, preserved in bogs around the lakes. Under the microscope, the pollen looks like fruit of all kinds, and this 'fruit' differs greatly from one plant species to the next. The masses of identifiable plant pollen Selling found in our samples completely altered the historic image of the Easter Island landscape.

Easter Island had once been wooded. Like Mangareva and the nearest island in Polynesia. Numerous types of pollen from palms and other trees, unknown on the island today, were found buried in the deeper deposits of the bogs. Even pollen from a palm and a coniferous tree not known from any other island in the Pacific. Above these original levels was a layer that contained ashes from extensive forest fires, marking the arrival of man. In the layers above this, the original species vanished, and only pollen from grass and ferns continued to appear, mixed with those of imported species. Among them were two aquatic plants that survived all subsequent fires because they grew in the lakes: the South

American *totora* reed and a South American medicinal plant.

The forest fires could not have been caused by volcanic activity, as the ashes blew into the lakes at a time the craters were extinct and already filled with water and plant plankton. Hence the fires signaled man's presence. They must have ravaged the landscape after the first seafarers had discovered the island and begun clearing temple yards for stone platforms with enormous monuments around the coast.

The people who came were stone masons, not wood-carvers. They cleared the woods in order to open quarries for their megalithic works, for villages of peculiar circular stone houses, for vast fields of South American sweet potatoes.

The discovery that Easter Island was as wooded when the first seafarers arrived as were the islands further west, dealt a fatal blow to the widely accepted dogma on which most theories on the origin of Easter Island had been founded. It was held that Polynesian fishermen 'blown out of course' had landed on Easter Island, as no South American stone-carvers could have managed a raft voyage into the open ocean. And so, stranded on a lonely island with no wood but plenty of stone, the fishermen gave up wood-carving and got the idea of carving men out of the local rock, gradually making them larger and larger until they were mass-producing monuments surpassing even those of the pre-Incans in South America.

Traditional histories of Easter Island state in plain words that the landscape was formerly wooded, even with timber of large dimensions. These tribal memories are also full of references to a period of civil war antedating the European arrival. While 'Long-Ears' and 'Short-Ears' fought each other, the burning of houses and property was a commonplace way of exacting revenge. It is therefore difficult to judge how much of the forest was intentionally cleared away and how much was lost when fires burned out of control. But for generations the descendants of the early settlers have paid dearly for their careless behavior toward their own environment. Wood has for centuries been so scarce that it has been hoarded as precious material, to be used for carving sacred written tablets and for breastplates and other ornaments and insignias for people of high rank. On Easter Island there is no shade from the burning sun, except inside houses and caves. On the whole island there is not a single rivulet. Rain sinks straight into the ground for the lack of forest roots, while traces of dry beds from former streams can be seen on the bare rocks. The only water supply comes from three stagnant crater lakes. Early visitors saw how Easter Islanders were diving along the coast with gourd containers held upside down to fetch brackish water that emerged below the surface in places where their own ancestors could drink from running streams.

At enormous costs of time and labor, the island is now being replanted with small groves of coconut palms and clusters of eucalyptus trees set far apart. Efforts are being made to reintroduce to the island the endemic *toromiro* tree, using seeds our expedition rescued from the last tree, a tree that died out on the island in 1956. But with millennia of accumulated humus washed into the sea, man will never again be able to reshape Easter Island as nature had created it, with all its original resources, running streams, and natural beauty.

Many years would pass before I would come to realize that the fate of Motane and Easter Island is sneaking up on us almost everywhere on our planet. In a few cases we get temporarily shocked and wake up in anger, as when Mururoa Island, west of Easter Island, gets bombed and forever contaminated by French nuclear bomb tests. But where the destruction advances slowly over the centuries we do not even see it. We assume that our planet was presented to man the way we see it today. That the Sahara was always a vast desert. That parts of Greece and Spain always presented a bleak and barren landscape.

I was to learn what we all ought to know: that while we look for treasures on other planets we let the one we live on slowly decay. I could not escape such meditations after I had traveled deep into the Sahara to look for man's earliest paintings of water craft. At the naked rock plateau of Tassili, 2,000 meters above the sea and about 1,000 kilometers from the coast, deep into the heart of the Algerian desert, the former inhabitants of the Sahara paddled reed boats to hunt hippopotamus on rivers and lakes. On smooth, overhanging rock faces between five and six thousand years ago, they painted in elegant lines the incidents of their daily lives when they hunted all sorts of forest animals and herded domesticated flocks of cattle. At the base of the cliffs, over a bone-dry waterhole in the rocks, they sculptured the relief of a weeping cow, tears falling from her eyes as she bends down to look for the water that has forever disappeared.

What happened to North Africa? Why did waterways and fertile pastures dry up to become the world's largest desert? This desert still advances southward and eats up central African jungles with a speed measurable in kilometers per year. Classical writers from the centuries around the time of Christ tell us that when the Greek and Roman conquerors reached North Africa, they found a green continent abounding in timber and fertile soil. The modest grain cultivation by the aboriginal Berber population was quickly expanded at this time, until North Africa became the fertile granary of the Roman Empire. The Berbers had navigated in reed boats, but the Phoenicians who had settled among them and founded commercial city-states, began cutting timber for wooden ships. This modest exploitation of the forests increased when the Romans needed timber for their naval fleets. The Arab armies later conquered all former Berber lands, and in the wake of the soldiers came the shepherds, moving into North Africa to slash-and-burn and to raise ever-increasing herds of sheep and goats. In the end, man's failure to protect what nature had offered him enabled the climatic change that had already begun about

3000 B.C. to turn jungle humus and cultivated soil into sterile sand. What Roman mosaics from the time of Christ depicted as a colorful paradise was transformed into the Sahara's endless desert dunes. The ancient Greek historian Herodotus wrote that in his day one could travel from Tangiers in Morocco to Alexandria in Egypt along the entire coast of North Africa – in the shade of green foliage.

In my search for the beginning of ships and navigation I have also struggled on foot through the hot sand of the Nubian desert between the Nile and the Red Sea. A more godforsaken landscape can hardly be found, even on the moon. Not a straw. Not a bird. The sun roasted the rocks until they split. No shadows but our own. Yet here too I had come to look for pictures of ships that had been incised on boulders and mountain walls before the first Pharaoh was born. I walked through a dry canyon that had once been a river. Here were petroglyphs of waterbuck, antelope, giraffe, elephant, lion, crocodile, and in the midst of such animals, which are no longer found within a thousand miles, were ships. Large sickle-shaped, reed ships with mast, sail, and cabin. Some even had two cabins on deck, and some were depicted with horned cattle on board.

What had happened here to the lost waterways of people who sailed them with livestock aboard before the first pyramid was built? What about their green hunting grounds?

And what happened to the biblical paradise of Mesopotamia, placed by Jews, Christians, and Muslims alike in one and the same place: in Iraq, where the rivers Euphrates and Tigris meet. There a desert traveler will find a plaque inscribed: THE GARDEN OF EDEN.

Believe it if you wish. The two rivers exist today as mere drainage canals from mountains in distant Turkey and through the endless Mesopotamian desert until they empty in the lifeless mudflats of the Persian Gulf. A Sumerian poet who lived here while the pharaohs reigned in the Nile Valley, wrote:

'To Ur he came, Enki, king of the abyss ... O city, well supplied, washed by much water ... green like the mountain, *Hashur*-forest, wide shade ...'

This was written about the arrival in Ur of the first Sumerian god-king, who came to Mesopotamia by sea in a *magur*, a large, sickle-shaped reed ship. There is no shade in Ur today. Not a green straw, not a drop of water; even the river has changed course and floats far away from the ruins of Ur's colossal pyramid. The houses from Abraham's time are totally buried in sand.

The wastelands housing the mythical Garden of Eden show that Adam's descendants were busy on the seventh day. On ceramic seals and temple walls excavated from the sand dunes are reliefs showing forest trees, game, and what appear to be tree plantations, and also of fleets of river barges loaded with high stacks of timber and the colossal logs that were used in the levering and transporting of megalithic stones and giant statues. Charred timber of a pine species unknown in Iraq today has been excavated from the ruins of Niniveh, where temple reliefs depict a fierce naval battle with men and women fleeing in reed ships. Overexploitation, arson, and mismanagement of the rich environment put an end to man's happy Sunday in Eden.

Once upon a time, the mighty cedar forests of Lebanon welded the Phoenician and Egyptian nations together. A special port in the harbor of Byblos was reserved for regular maritime traffic carrying timber to temples and shipyards along the Nile. But when the timber-greedy Greeks and Romans conquered Asia Minor, the fabulous cedar forests came to an end. Apart from the one green tree proudly preserved on Lebanon's flag, there are only a few clusters of cedars left as a national monument on the stripped hills of Lebanon.

By the fourth century before Christ, the 'green' movement started in Greece, with Plato and his pupil Aristotle observing the interaction between deforestation, water shrinkage, and soil erosion. 'Our land,' Plato wrote in one of his dialogues, 'compared to what it was, is like the skeleton of a body wasted by disease. The plump soft parts have vanished, and all that remains is the bare carcass.'

At the time of the Persian War, Athens was so dependant on its naval strength to survive that Sparta succeeded in strangling them in 404 B.C., first by blockading the importation of shipbuilding timber from Macedonia, and then by destroying the supply of logs Athens had stored in Italy.

Today Greek authorities try to arrest the forest destruction, but, man having opened the way, an estimated 8 million goats and 16 million sheep continue the decimation. A report to the International Union of Forestry Research Organizations suggests that in a recent 20-year period, 75 million cubic meters of Greek soil were washed into the sea, decreasing the productive area by 375,000 acres. An amount of land the size of the entire island of Rhodes.

The foundation for global civilization was shipbuilding, and the art of plank-built ships spread with Afro-Asiatic civilization from the Middle East to Greece, and from there to the Italian peninsula and the rest of Europe. Deforestation of the Atlantic nations started on the Iberian peninsula, where by the end of the medieval period the Castilian landscape was already so denuded that there was an old saying that a bird could fly over it and never find a branch on which to rest.

Through unlimited exploitation and expanding farmlands the forests were felled northward through the Low Countries, and the large oak forests of Norway available to shipbuilders in Viking times were gone by the eighteenth century.

With the coming of the Spanish conquistadores, deforestation spread to the New World. With ax and fire, European pioneers began their unrestricted war against the jungles of Brazil, where the flames are still blazing with undiminished fury. The vast demand for timber to be used as mine props played an important part in land denudation from the Mexican highlands to the

Andes. Mexico was conquered by 1521, and as early as 1543 the Indians of Taxco complained to the viceroy that all the forests within a day's travel from the mines had been cut down. Botanists have estimated that 75–80 percent of Mexico's forests must have been destroyed since the arrival of the Europeans.

In what today is the United States, the first European colonists found the forests to be so extensive that it was said a squirrel could travel from the Atlantic coast to the Mississippi River without touching the ground. The pioneers complained that the trees stood too thick for their wagons to pass, and when they cleared the forest, they often piled up logs and burned them because they had more than they needed. Wildfires today destroy about 5 million acres of forest each year in the United States – more than the sum total of all Britain's woodlands. Because nature is being brought out of balance, insects and disease tripled the destruction of United States timber during the first half of this century and today destroy more than 1.5 billion cubic feet annually.

The firewood crisis is mounting and has become acute in densely populated parts of India, in the increasingly arid belt of Africa south of the Sahara, in the Andean region, in deforested parts of Central America, and in the Caribbean. In some areas people have begun to rake the hillsides clean of layers of leaves, which they use as home fuel; this has become a principal reason for soil erosion in Asia, where all rivers carry immeasurable volumes of badly needed humus into the Indian Ocean.

Many years after my unexpected encounter with the doomsday landscape of Motane, I began my own case study of what happened to Planet Earth in the corner of the world where I was then living. The environment was changing in front of my eyes from year to year. I had settled in the most fertile and beautiful landscape of the Italian Riviera, in a medieval hamlet perched high on a ridge overlooking the Mediterranean, terraced by olive plantations and surrounded by densely forest-covered hills and valleys. Behind rose the Alps and, sheltered from the northern winds, in front lay the sea that brought culture to Europe.

The Romans had built the stone-paved Via Aurelia that passed through my property. Napoleon had used it to bring guns and men eastward to Austria, and when Pope Pius VII passed there on his way from Rome to Avignon in 1814, he descended from his litter and blessed the local farmers assembled in the little chapel that stood on my land. For two thousand years, generations of travelers had seen the local landscape changing. But never with the speed that was measurable in the latter decades of the twentieth century.

The bulk of the coastal forests of Italy and France were gone within a single generation. The flow of tourists swarming to enjoy the local beaches after World War II resulted in the farmers leaving their terrace cultivation and moving to town to cater to the tourists. Dry grass and underbrush replaced the former fields and pastures, and careless people set the forests on fire. There were no longer deer or other forest game to come back and clear the undergrowth, which for many centuries had been kept open by domesticated livestock. The trails disappeared and the forests became impenetrable for man, but not for the roaring, blazing flames that devoured thorns and timber alike.

A team of Swedish soil geographers joined me as we mapped all forest fires that had raged in the adjacent municipalities over the last twenty years. There were no longer any streams within the area, but we placed filters in a number of dry ravines that carried water for a day or two after heavy rain. We discovered that we could drink the clear water that ran down into the valley through a terrain that had never burned. After one or two fires had swept through, the water was coffee-colored, and there was mud in the filters. Where three or more fires had been registered, all trees were gone. Fires had followed the dead roots and charred the humus so that only sparse grass was left. In one such test area, the soil surface sank a good ten centimeters after a heavy rain. Calculation based on the runoff revealed that a tiny creek, dry except after rain, would carry as much as sixteen tons of fertile mud down the slopes and into the sea. Slowly and certainly this Riviera follows in the footsteps of northern Africa. Crete. And Greece. Our only hope in stemming the complete deforestation is the 'green tide' among local youth, who have started voluntary forest-fire brigades. The fires can be stemmed. But what can members of the 'green tide' do elsewhere in Europe, where they see their forests die from fires they can never quench? Fires from industrial fuel burning in cities and countries far from their own? Rain loaded with sulfuric acid from tall chimneys in England and continental Europe fall over Scandinavia and reduce the growth of trees and kill the fish in the crystal-clear Norwegian mountain lakes. First one, two, a dozen lakes got so acid that every single fish ceased swimming about. We at the university were shocked when we heard about it before I left for Fato-Hiva. Now the dead lakes in Norway are counted by the thousands, and diplomatic notes go over the borders with diplomatic answers coming back: sorry, we can do nothing, national economy survives because of industry.

And industry survives.

XII

Across the Abyss to a Forgotten World

As the island of Motane dwindled and slowly sank in *Moana*'s wake, the stinging sea breeze seemed to wake me up from a voyage through time. I was longing to step ashore on the boulder beach of Omoa Valley and hurry in between the dense foliage under the jungle roof. On Fatu-Hiva our planet was still alive.

The tall palms seemed to be lined up along the queen's trail for the occasion. Not as stiff flagpoles flanking a royal parade, but as friendly personalities from the green crowd of Fatu-Hiva's familiar forest, welcoming us back to our own valley. We felt like waving back to them, for we knew them all, the tall ones and the short ones, and the crooked one that leaned across the trail before it realized its mistake and rose skyward like the rest. Every one of them was a familiar friend. And so were the nests of orchids riding on the gnarled branches above our heads. They seemed to be waiting for us, exactly as when we left. We knew also where lianas and aerial roots hung down from the tunnels of greenery. We greeted them in passing, touching them and leaving them dangling behind us.

This was our world. This was our valley. We were back where we belonged, heading once more for our yellow bamboo home. We were happy.

Barefoot again, and wrapped in our light pareus, we felt anew the provocative touch of life, as nature played on us with all its instruments to keep our senses wide awake. The jungle caressed our skin and filled our lungs with a warm fragrance. We were no longer on a dead planet. Motane had been left behind, as if it had sunk into the seas. We were re-entering a living world, where the birds and butterflies still survived to flutter at will in the sun or in the soothing shade of the jungle. Here a deep padding of rich humus still covered arid subsoil and gave birth and a foothold to spindly mushroom legs, tender stalks, and giant trunks. Here was no sterile sand, but a soil teeming with growing and moving creatures. This was a sample of mankind's fabulous heritage: nature on the seventh day.

We hurried inland, gay and filled with happy expectations, like children going to a birthday party. Once more, all this was ours to dive into afresh, to grasp with all our appetite and our senses. It was densely overcast, yet the sun seemed to beam from above, the weather independent of the clouds. The jungle air was as raw as when we had last inhaled it, but now we enjoyed any form of humidity. We even enjoyed the musty scent of mold and other fungi. All combined to reflect the abundance of fertility and life. On the day before, we had seen an island where nothing as worldly as fungi would grow. Overnight, we had made the leap from an extinct world back to a living planet. A planet where pearls of water still dripped from the leaves. Where chuckling rivulets tiptoed into dancing streams. And where busy insects scurried about among the greenery, searching for proper places into which to poke their feelers, as if they had been paid by their bosses to grease the complicated machinery that grew about us and gave us shade, air, and food.

As we gazed around us, the scales again fell from our eyes, and we marveled at everything like globe-trotting discoverers, although we were approaching nothing but our own former doorstep. Yet today, once more, nothing was obvious, and all we saw in nature appeared worthy of a second thought. The barrenness of Motane was still fixed in our mind's eye. Motane was, after untold millennia, returning to the primeval dead condition of all these islands. Fatu-Hiva, too, had originally been just as naked. This luxuriant forest was here today. Not yesterday. And tomorrow it might disappear again. It was a loan exhibition for an undetermined time.

Where had all these millions of tons of sap-filled timber come from, this complicated mass of interdependent life? A few wind- and waterborne seeds had arrived here after the blazing volcanoes rose from the sea, their minute cells loaded with all that was needed to explode into this profusion of living, dying, rotting, and reconverted species. Neither man nor any living beasts were scared or surprised at this magic transformation; it happened so slowly as to be imperceptible to any eye. Time hides from the brain what the eye should register as a miracle. Man, like the beasts, is designed to react to sudden changes in the environment in his own self-defense. We do not see the ever-present tip of our noses, nor do we hear the constant drone from a street or a waterfall. A dog will sleep under an apple tree and a bird perch peacefully among its branches, and neither will notice that the tree is alive and the apple growing until the fruit suddenly falls to the ground. Permanence and very slow motion put us off guard.

But the jungle rose too quickly over our heads the day the schooner brought us back from naked Motane to lush Fatu-Hiva. We were out of tune with time. We felt what we knew; the valley we now entered had been more barren than what Motane was today, without one green leaf, without a single shovelful of humus. Time had hidden from all eyes that this jungle scenery had been brought from nowhere, came from nothing. The barren rock of Fatu-Hiva had been lifted up from the ocean floor like an empty picnic table ready to be laid with fruit and nuts, and decorated with orchids and other flowers. Behind the greenery we sensed another landscape, Fatu-Hiva as it had once appeared, naked and newborn as a baby, while early man still slept in the chromosomes of his primitive predecessors. The air full of smoke and steam, red lava oozing down all the ridges and ravines, while glowing cinders and ashes rained into the valley bottom that would later hold a forest and a river filled with prawns and drinkable water.

When we reached this river, we sat down and enjoyed a pleasant temperature perfectly adjusted to our human bodies. The volcanoes were gone. We sat down on a wide, smooth slab to look at the running water and enjoy its melodious, nonstop performance. The only other musicians were some twittering finches on a moss-covered trunk, who seemed to tune in perfectly with the river. At the crossing place, the water streamed as rapidly as it had during the heavy rain before we left. The soil was still soft and moist. A big white mushroom had come up from the mud beside the tree trunk. It had grown so fast and determinedly that a stone had been lifted aside, together with a lump of clay. I touched it with my big toe. It fell over. So fragile, and yet able to push its pale head up through all obstacles. Freshly risen from the dirt, this visitor from the underground had a white hat and a white neck as shiny and clean as if it had never touched the soil from which it was created.

The fallen mushroom exposed the lips of its fertility organ under its skirt: a dense system of blades radiating around the leg. A few days ago this fair lady was the size of one of her own tiny pores. Perhaps tomorrow she would stop growing. What stops the mushroom from growing bigger than its allotted measure? What prevents it from reaching the size of an umbrella, or a circus tent? There were tons of deep, rich soil to permit growth; what prevented it? Why had none of Fatu-Hiva's wild, lusty species managed to upset the balance around us, conquer the island for its own kind, and transform the valley to a wasteland like Motane?

I was fresh from the university and could give the answer to Liv that I knew my professors would have given me: the chromosomes were able to keep order and prevent chaos. The living cells were able to divide themselves into a hundred billion cells of all sorts, all guided in the right direction by the chromosomes. When the intended organs and the correct form and size of a growing creature have been reached, the program of the chromosome has been fulfilled and the cells automatically stop dividing. Each one of the billions of independently dividing cells will end its fabrication process the moment the combined task is accomplished; when horns and intestines, teeth and tails, ears and eyes are ready to operate together, the chromosomes have done its duty.

Damn the mosquitoes! They had begun to accumulate in clouds around us, and we both rose to put on clothes. We wished the chromosomes could have been less successful in composing the long noses of these flying devils.

We picked up our light burdens and returned to the trail. To our delight, we found ripe guavas on the bushes along the path. There had been much rain but little traffic up the valley since we left. Of course: Willy and Ioane had come back with rice and flour.

We approached our clearing on the royal site. Liv was thirsty and turned to drink at the queen's spring. As she stopped and pushed aside the umbrella plants, a *poto* or wild cat ran out from among the big leaves.

Nearby, the yellow plaiting of our bamboo cabin was visible through the greenery. I loved this place. This valley. I had never been as happy as in this hospitable wilderness. Relaxed between all the elements of the green foliage. Secure in familiar surroundings as part of the perpetual clockwork where nature still was the master. But we had to be alert to the diseases unknown in these surroundings before the caravels and schooners arrived.

There was the cabin. Beside it, the unwalled kitchen. The ground was still quite muddy. To my delight, no footprints.

But what a change!

Liv was now at my side, and with a cloud of mosquitoes around us, we hurried up to the front wall, which was partly hidden by huge, sprouting banana leaves and other fresh vegetation. The jungle had begun to recapture our clearing at an almost incredible speed. Everything seemed to have bounced straight out of the ground in the moment our attention had been turned elsewhere. Had we been away a short time or a year?

An exclamation from Liv told me that she too had noticed what I was gazing at. The four poles I had sharpened and rammed into the ground to support the kitchen roof had sprouted long branches with green leaves.

I tried to open the door of the cabin. The whole frame yielded. Surprised, I tested the strength of the beautifully plaited golden wall. My fist went through it as if it were thick paper. I peeped inside. Tatters hung down everywhere from a caved-in roof. Spiders and centipedes ran up the walls as we entered. And once more, the bamboo dust. It poured down upon us if we touched the feeble walls. Like a coating of snow or desert sand, it had covered our homemade furniture, our sleeping bench, and all we had hidden under it: stone tools, images, and skulls.

Nobody had been in. Our archaeological and zoological collections were safe. But we ourselves had no home now but the jungle roof.

We pulled out from the wreckage of the cabin the case of craniums and all the other property that we did not want anybody to find when the walls collapsed completely. We hid it in a dry cache among the rocks. We praised our good luck that the rain was not pouring down, and hurried to raise a crude shelter of sticks and large umbrella leaves on the clearing in front of our former home. With our old mosquito net carefully stretched over us, we fell asleep on a thick mattress of ferns. Then the rain started to drum against the dark world around us.

We were wet when the morning came. The rain had stopped, and mud and mosquitoes surrounded us. This was, after all, not a place to remain in. We could build a new bamboo cabin on the site of the old one, but the same things would happen again. Besides, it was too risky to remain here in the mud. We would get the *fe-fe* on our legs all over again, and any one of these mosquitoes could carry the filaria of elephantiasis.

'We shall have to do what the islanders do,' Liv suggested. 'Abandon the jungle and live at the coast, where the wind drives the mosquitoes away.' I agreed, but then we must escape to some other valley. There was so much disease among the islanders that we could not live with them in the village down by the bay.

We resorted to Pakeekee and told him our problem, but he merely insisted that his home was ours too. Tioti seemed to read our minds. If we wanted to escape from the mosquitoes, he said, we should try to get across the Tauaouoho mountain ridge. On the other side of the island, the constant wind from the east always blew so strongly that the insects were chased far up into the thickets of all the valleys.

Neither Tioti nor Pakeekee had ever been across the mountains to the other coast, but Veo had. He confirmed what we already knew, that the Tauaouoho and Namana mountains join to form a mighty wall that separates the windy, east-side valleys from those of our sheltered western coast. Since the needle hole through the ridge between Hanavave and Hanahoua was no longer accessible, the only passage for man was at one specific point on the central plateau, where a very ancient trail had once been cut into the cliffs leading down to Ouia. Ouia was the largest valley on the other coast. Landslides had carried away much of that passage too, but with care people could still get down.

Except for Ouia, the other side was uninhabited. All the tribes had died out. All the valleys were empty of people ... except ... In Ouia lived an old man, called Tei Tetua, all alone with his adopted daughter. Veo had met the old man. Tei Tetua had once been the chief of four tribes, but had survived all his people and all his twelve wives. A relative in Omoa had brought him Tahia-Momo, Little Tahia, the young girl who kept him company.

'Tei Tetua is the last of the men from the past,' Tioti explained, and Veo nodded. The old man belonged to the ancestors. He was the only one left of those who had eaten human flesh.

We knew that cannibalism had been practiced here at least until fifty years before we came. A Swedish carpenter had been eaten on Hivaoa in 1879. The last recorded instance of cannibalism on that island was in 1887, during a ceremony in the Puamau Valley. Tei Tetua was a grown man then, and his valleys were far more isolated than any part of Hivaoa. But even an ex-cannibal would long for company when left alone. At any rate, there was no choice of valleys on the other coast, for, according to our friends, Ouia was the only one we could reach on foot. From there, one could get nowhere, as all the valleys on that side were isolated from each other by insurmountable cliff walls.

Veo could also tell us that the other side of the island was much drier. The clouds that build up over the Tauaouoho's ridge blew over onto this side before they dropped their rain. There were fewer mosquitoes, too, on the other side.

Tioti, and Pakeekee's agile son Paho, both volunteered to

follow us across the mountain ridge. But not Veo. And he was the only one of them, the only one of all our friends, who knew the road. Nothing, not even a supply of gifts we had bought on Hivaoa, could tempt him. Nor would he give us any reason.

Paho suddenly appeared out of nowhere in front of our ruined bamboo cabin. If we gave him the gifts from Hivaoa he would share it with a man who knew the trail across the mountain pass. He would not tell his name; nobody ought to know it.

After dark we followed Paho with all our possessions down to Pakeekee's hut, where we went to sleep on a plaited pandanus mat on the floor.

It was still pitch black with no sign of sunrise when Pakeekee called us. The village slept, and all we could hear was, at fixed intervals, the growling of the surf. Our friends had already loaded our few possessions on two horses, and we threw our blankets on top.

Tioti warned us that nobody should know where we had moved to. He trusted no one. We whispered a farewell to Pakeekee and his hospitable household, and with Tioti, Paho, and the two packhorses at our heels, we walked silently down to the bay. A few dogs barked lazily, but we saw no one.

Where the familiar trail took off for the highlands, we found our third companion, a young man we had seen before but did not know by name. He waited till our caravan had passed and joined us at the rear.

The sun did not rise that morning, but the clouds turned from black to gray, and we gradually obtained good visibility as we followed the winding path through the glens and hills we knew so well in the central highlands. Once up there, our guide finally took the lead, and after some hesitation, located an overgrown trail that branched off from the main path and took us due eastward, in the direction of the highest peaks. The route we followed was often swampy and difficult. For a while it led through a mountain forest, where trunks of fallen trees occasionally barred all passage ahead. In the afternoon, we reached some bamboo thickets that blocked the road like bulwarks. We had to cut a passage, step by step, with machetes. Bamboo canes, thick and thin, yellow and green, were interlocked in a crisscross chaos where, once the canes had been chopped off, their knifelike edges pointed against us like drawn bayonets.

Tioti hurt himself rather badly when a thick bamboo pole slid down as he cut it, and speared his fist. As Liv sat down to attend to his bleeding wound with leaves and bands of tough bark, she caught a glimpse of his right ankle and was shocked. Without interrupting the bandaging she gave me a silent sign with her eyes, and I looked. Tioti had begun to wear long pants but with one leg split so as to give room for a colossal foot and ankle. His right leg had started to enlarge with elephantiasis, and he was trying to hide it. We both were overwhelmed by this tragic discovery as we rose and continued eastward at the heels of the cheerful sexton, Tioti. And we realized more than ever how important it was to escape the clouds of infected mosquitoes.

The east wind struck us full strength as we reached the point where the cliff edge fell off in a vertical drop, down into the deep abyss of another underworld: Ouia. For the horses, there was no longer a foothold and we had to tether them, each to its own tree, where they could stay grazing while we proceeded on foot. Then our three companions cut carrying poles and loaded the burdens onto their own naked shoulders. At this point, we had to start climbing. The wall was no less than perpendicular, but in ancient times men had carved a narrow shelf toward the left, on which it was possible to get a foothold. At several points, the old shelf had been destroyed by erosion and slides, but our guide was prepared and had brought a strong pole to lay across the open gaps.

We were both admittedly frightened. But we had little choice. The schooner had left the moment it had set us ashore; up here in the highlands there was not enough food, and our home in Omoa had been conquered by bugs and mosquitoes. It was clear to us that we had to overcome our desperate feeling of dizziness and get off this cliff. We started the climb down.

Before the descent began, our three companions were visibly excited at the prospect of meeting one of their own kinsmen who had tasted human flesh. Cannibalism was considered the most horrible sin of their pagan ancestors, but it seemed to be a common opinion that they had sinned because they had eaten *unclean* people, and thus got inherited evilness in their own blood. For at that time all people were pagan and unclean. But if a person was clean, like Christ, they said, one could eat him without getting defiled. Tioti, as a Protestant sexton, had a hard time trying to convince the two others, even Pakeekee's son, that the consumption of Christ in church was only symbolic, a way of showing that they liked him.

Down in the deep, dark valley, with the vertical abyss rising around us on three sides, we followed the course of a gushing river through a jumble of crooked hibiscus trees. The ground was covered, moreover, with a riot of boulders, and we had to proceed as best we could without any kind of trail. Halfway down, the entire river suddenly disappeared into the ground. All the gushing masses of water were lost below the terrain. There were more stones than mud as we proceeded. The river suddenly reappeared down by the mouth of the valley, welling up from among the rocks, and we followed it as it danced on between boulders toward the sea.

Paho had run ahead along the river, and the distant baying of dogs told us that he had reached the old man's hut. We had not far to go.

The valley had gradually become much wider, and soon it became bright and open. The thick brush yielded to a beautiful palm grove, down by the glittering sea. The sun broke through. We inhaled a fresh wind from the open ocean. To the right, by the beach, we saw a cluster of low, sunbaked huts between lofty palm trunks, built in the old-fashioned Polynesian way from sticks and thatch. A seemingly naked person came running toward us between the trees. The old man.

Tei Tetua ran like a young mountain goat. He was weather-beaten and suntanned all over and wore nothing but a sort of penis bag suspended from a band of bark around his waist. The old man was muscular and agile, and looked as if he were half his real age. His whole face seemed to grin as he laughed with happy animation, showing teeth as perfect as those of the ancient skull under our bed. When I gave him my hand and said *kaoha*, he grabbed it, laughing and writhing like a shy boy short of words. His whole person was almost bursting with restrained energy. He just did not seem to be able to express all he had to say after years of loneliness.

'Eat pig!' he finally exclaimed. 'When pig is finished, we eat cock. When cock is finished, we eat more pig!'

XIII

In the Cannibal Valley

TEI TETUA BOUNCED AWAY like a boy, down to his cluster of huts, and began shouting at his bristly, semiwild boars. With the aid of Paho, he got a bark lasso looped around the hind leg of one of them and came staggering toward us with the huge black beast struggling in his embrace.

'Eat pig,' he said merrily, and showed us the fat and shrieking hog.

To him, that was clearly the greatest expression of friendship. And until late at night, we sat on his earthen floor eating juicy boar baked underground. We squatted around a sparking wood fire with large chunks of hot meat in our hands, and devoured it greedily with teeth and fingers. Close to the old man's side sat Tahia-Momo, his adopted daughter. Barely a teenager, she was beautiful. With her large, shining eyes and long, black hair, she sat squatting like the rest of us, listening attentively to every word that was said. Tei Tetua, too, was as alert as a child, beaming with happiness at seeing people in his lonely valley.

'Remain here,' he asked all of us insistently. 'Ouia has much fruit. Much pig. Every day, we shall eat pig. The wind is good in Ouia.'

Liv and I promised to remain in his valley, and old Tei and little Momo laughed with joy and thought up the most tempting plans for the days ahead. But our three friends from Omoa shook their heads. They wanted to start the journey back next day.

'Ouia is not good,' said the sexton. 'Ouia has much fruit, much pig, much wind. But Ouia has no copra, no money. Omoa is good. Many houses, many men. Much copra, much money.'

'Tioti,' I interrupted, 'what do you want money for, when you can get all the food you need without money?'

The sexton grinned. 'It is like that,' he said and shrugged his shoulders. 'Before, it was good enough without money. Not now. We are no longer savages.'

The old man knocked out the fire and scooped ashes over the embers. Liv and I were shown from his open kitchen shelter into his stick-walled hut, where he left his own mat on the earthen floor for us to sleep on. He himself moved out to sleep with all the others in a separate hut beside our own. We were never to learn the original function of that hut. But the closely set sticks that formed the walls of both Tei Tetua's dwellings were gray with age and of a very hard wood that might well have dated back to the days when he had other company than little Tahia-Momo.

We rolled ourselves up in our mat. How marvelous it was to go to bed without mosquitoes! The surf was so near us, somewhere just beyond the airy walls. In Omoa we had been used to hearing the distant breath of the ocean carried to us on calm nights above the jungle roof as a faint, rhythmic hiss. Here, the boulder-beach surf rumbled and snored with the same slow rhythm, but with a strength that made it seem as if we shared a pillow with the ocean. It would take time for jungle dwellers to get used to sleep with the ocean this close.

The others remained squatting around the embers, stirring their glow and talking in low voices so as not to disturb us. Tioti repeatedly mumbled our names and was probably telling our story. Old Tei had a long, exciting tale that seemed to keep the others spellbound. Several times we heard the words kaikai enata, eating people, and were sure our friends were pumping our host for stories about cannibalism.

Kaikai enata. As I was slowly falling asleep to the lullaby of the sea, I probably heard that phrase more often than it was spoken. We had constantly added to our little homemade dictionary, and although our knowledge of the Marquesan dialect was still rudimentary, we were familiar with most everyday words. Liv lay motionless but wide awake beside me. I could not help thinking of my father-in-law, who had run to his bookshelf and read about cannibal islands when Liv wrote to him about our plans. What would he have said now, if he had known we were lying on the mat of a formerly practicing cannibal, with the host squatting in the dark outside the wall. If this cabin was only as old as my father-in-law, somebody would probably have been eaten within these walls.

The sexton seemed to be in no hurry to sleep, and each time he poked the ashes to get more light, a flickering flame created dancing shadows in our room, for there were as many open cracks as there were crooked bars in the walls. In the flickering light we noticed some large brown gourd containers suspended from wooden hooks on the walls, and some bowls of coconut shell, black with oil, elaborately carved in geometric and symbolic patterns. A bundle of dry tobacco leaves hung in a corner. Some old stone adzes and a crude iron hatchet had been flung down beside us on the earthen floor, together with the old man's flint and steel. And down from the thick beams of the roof hung a long and strange wooden box. That was Tei Tetua's own coffin.

'If I get ill,' the old man had told us quite cheerfully, 'I shall just crawl into the coffin and shut the lid. If I should remain on the floor, the dogs would just break in and help themselves.'

The old man had also dug his own tomb beside the house. He had put a gabled slab roof and a cross over it, and later we often found him inside throwing out dirt kicked into it by his chickens.

This was a new world to us. No schooner came here. At last, we were as far away from the grip of civilization as we could possibly get.

The embers died out. We fell asleep and no longer heard the thunder of the surf. Our friends in the kitchen shed, too, must have withdrawn to their hut. They had to get up before the sun the following morning and start their long return trip to the west coast.

We woke to find ourselves alone with Tei Tetua and little Tahia-Momo. The sun shone, the birds sang. The others were gone. The valley was as open and hospitable as the day before, and our two new acquaintances as happy and smiling. This was the land of our dreams. This was where we should have come from the very beginning.

Tei Tetua was sole proprietor in the Ouia Valley – at least, there were none to contest his rights, and we were free to choose any site we wanted without paying rent. We were guests in the old man's kingdom and anything that was his was ours. Tei Tetua had no feeling for personal property.

Tei's own cluster of huts lay in the open palm grove at the southern corner of the bay. An old terrace had been raised here at one time to keep the ground safe when the river flooded, and a solid stone wall was built all around it to keep out the wild as well as the semidomesticated boars that abounded in the depopulated valley. The shallow river passed right below the wall and entered the ocean through a wide passage in the huge boulder barricade that the powerful trade-wind surf had piled up along the curving mouth of the bay. This high boulder barricade sloped steeply into the agitated ocean, where the wild surf droned against it uninterruptedly and invited no one to swim or launch a canoe.

Across the river, a stone's throw from Tei Tetua's huts, and as close to the sea as they were, was a little grassy plain, which boars and wild goats seemed to favor. There was plenty of space between the lofty coconut palms, and we chose this site for our new home. Here there was a constant sea breeze and healthy, unpolluted surroundings, where insects and germs would leave us in peace. This was indeed the windward side. Chasing across the Pacific from South America, the eternal trade wind had not touched land for four thousand miles and they had jumped this high boulder barricade and made the trunks of the waving palms sway like grass, high above our heads. White man's stowaway devils, the mosquitoes, were swept far into the inland thickets. It was a sudden relief to experience Polynesia as it must have been before the unintentional introduction of these tiny vampires, which had come to make life unsupportable for us in Omoa.

The old man at first disapproved of our plan of sleeping anywhere but in his hut, but since we had picked a site so close to his, he yielded. We wanted to make sure that this time we built a home that did not collapse after a few months. He had been furious when, the night before, Tioti had told him of our bamboo cabin in Omoa, and his mouth was set hard when he thought of how his own people had changed. Nobody cared for the skills and knowledge of former days anymore. Now everybody just sat waiting for coconuts to fall so they could make copra and get something to eat from the schooner.

Tei Tetua's experience of modern life was rather meager, however. He had twice in his life been over to the other side, the last time to bring back Tahia-Momo when he got tired of being alone. Long ago, before everybody else had died, a missionary had come across the mountains and given him a proper baptism and a bronze cross to put on top of his tomb. Someone had also tried to land with a launch on the boulder beach to harvest coconuts from his valley, but the surf had wreaked havoc with the boat, and Ouia, like the rest of the cliff-girded windward coast, had remained unprofitable for modern man.

Tei Tetua was full of pride and enthusiasm when he spoke about people and events of long ago. In this respect, he contrasted with our good friend the departed sexton. Tei was in fact one of the extremely few islanders we had met who remained an unspoiled Polynesian in both body and mind. Like Teriieroo on Tahiti and Terai on Hivaoa. It took wisdom and a keen brain for one who had never been confronted with anything but the tempting side issues of our civilization to realize that progress is worth striving for only in those areas where progress means improvement. Not one of the disease-stricken islanders on the other coast, whose lives depended on the visits of the schooner, was half as carefree and contented as this old man who missed nothing, at least not now that he had company.

A climb at the heels of Tei Tetua took me the next day across rugged coastal shelves and around the northern promontory of the bay into a tiny valley called Hanativa, an easily reached addition to our own major valley. Old, overgrown stone walls, tombs, and a few large-eyed images were seen here, covered with a complete jumble of *borao* and *mio* trees. We had come to fetch *mio* wood, which the old man said was the best for building. I was completely amazed when I saw the natural agility with which this old islander, who could have been at least a big boy's grandfather, swung himself into the maze of trunks and branches. I followed with much less elegance. We were searching for the straightest possible sections for poles and beams. When a large quantity of branches of the desired thickness had been cut into convenient lengths, we pounded their juicy bark off with smooth stones, then tied the resultant slippery, ivory-like poles into bundles. My naked shoulder was rubbed sore and my bare feet cut by sharp lava as I climbed with my burden along the rocks as fast as I could. For the old man, all this seemed to be mere play, and he was audibly amused when I stopped at one point and clung to the rock wall as an outsized breaker drenched me in a runaway cascade.

As we reached the tip of the cape, we threw our bundles into the sea and returned for the next load. The surf carried the poles all the way into the bay and threw them up on the boulder barricade in front of our building site.

Our second home was destined to be even smaller than the first. It was in fact nothing but a palm-roofed nest, open on one side and raised on poles high above the ground to avoid visits from the many boars. The tree walls were made of sticks of *mio* wood lashed together, and were high enough for us to be able to sit anywhere, but to stand only under the central gable. The frame and everything else was tied together with tough strips of bark. The local species of hibiscus always provided a truly excellent rope. A sloping ladder led to the open side, and the interior space was just wide enough for us to sleep crosswise on a bedding of palm leaves piled up along the inner wall. Except for the waterproof palm roof, sun and moon could peep in everywhere through the chinks between the slender sticks, and small items kept dropping through the floor until we covered it with a freshly plaited pandanus mat.

During the first night in our new nest, we were awakened by a grunting boar. He was scratching himself so passionately against one of the main supporting poles that the top-heavy structure began to sway and threatened to capsize. In the morning, we strengthened our slender framework with a stockade that held all the supporting poles together and also prevented beasts from getting in underneath.

Our next project was to erect a kind of open kitchen shelter over a stone oven, like the one we had had in Omoa. But this caused the most vigorous protests from Tei Tetua. We were guests in his valley, and therefore we were to eat his food. He grabbed our black iron pot, which was the only kitchen equipment we had, and padded with his booty to his nearby house.

This was the overture to our happiest days in the South Seas.

Tahia-Momo – the old man called her only Momo, 'Little' – immediately took to Liv. To have another woman in the valley was for her as if a mother had descended from the lofty rock enclosures; a mother or a female companion, someone to talk to who had other interests and opinions than those of the old hermit. Momo would soon be a teenager, and was therefore considered almost an adult. If Momo had much to learn from Liv, she certainly repaid her lessons. Tei and I would frequently find the two girls sitting together in the grass beside the refreshing stream. Momo was the teacher, showing Liv the old arts once known by every Marquesan *vahine*: how to make *tapa* by soaking and beating the inner bark of the breadfruit tree into a fibrous white paper-cloth; how to weave artistic basketry from slender vines and coconut leaves; how to plait sleeping mats from thin strips taken from leaves of the pandanus tree; how to make strings and solid straps from the fibers of coconut husk; how to clean and roast the rind of bottle gourds to make them into waterproof containers; how to make glue from resins, and oil paint from earth, ashes, or vegetable components; how to extract perfume from plants; and how to prepare wreaths from flowers and necklaces from nuts and shells.

Liv never wore jewelry, but her little Polynesian friend resolutely undertook to dress her up. They did not have to go far before Momo found her choice. At the edge of the forest, she knew where to find colorful nuts, seeds, and fruit to thread on strings. Her favorite beads were some glossy red, bone-hard peas that looked just as if they had been designed as jewelry. With shining red necklaces and bracelets, and with perfumed flowers behind their ears, the two young women returned from the outskirts of the forest. Old Tei Tetua grinned and looked at me with an expression that implied that women all over the world had much in common.

Next day, Momo dragged Liv along in search of a new fashion. The shoreline of rugged lava running seaward below the cliffs on each side of the bay was to Momo a regular shopping center, displaying small seashells of every shape and color, all for the picking. Their green shopping bags filled with marine jewelry, the two artists returned to the soft grass to pierce their shells with bone needles and thread them on strings.

There was beauty on display in all Momo's shopping centers. No evil-smelling copra was needed for barter; she had a fortune in her bare hands. We never saw Momo's face without a smile. She knew what she wanted, she had a certain taste, and she never complained about anything she could not get. Her taste and harmony must have come to her from the hills and trees, hardly from the old man. She seemed to have no problems about herself or her environment. She laughed when she came climbing up our ladder to visit us and she laughed when she left. She saw something pleasant in everything, even in the little gray creepy-crawlies hiding beneath the stones.

Tons of smelly copra carried to the schooner on the other side would not make the women of Omoa and Hanavave radiate happiness like little Momo, even if it had sufficed to give them access to all the glass and metal jewelry in the shop on board.

There were times when Liv was too conservative to follow some of Momo's island fashions. One afternoon, for instance, the girl came climbing up our ladder, her white teeth laughing out of a yellow-green face, and with yellow-green paint shining all over her body. She carried a coconut shell in her hand, filled with the greenish pulp from a nut pounded in coconut oil, and she wanted to paint her friend with genuine Marquesan cosmetics. How creative! Yellow-green and with flowers in her hair, Momo looked like a charming fairy-tale elf born of a cabbage. The effect was not altogether disgusting and might well have caught on in a more modern society. But Liv felt she should stick to her more conservative use of pure coconut oil, to which she let Momo add the sweet perfume of tiny white flowers.

As for myself, I spent most of my time with Tei Tetua, in the forest or by the huts. If old Tei had wanted to come and live with me in Europe, I could have taught him many lessons that would have improved his own existence. But I was not the right person to better his existence here. In this environment, I was the one to benefit from learning rather than from teaching. I could tell Tei the Latin names of some of the mollusks along the waterfront, and how to classify plants by counting their stamens, but I found it more important that he teach me whether the clams were edible and how the plants could best be used.

Perhaps, at the bottom of our hearts, Liv and I were a bit surprised at feeling that Tei and Momo were no closer to the apemen than we were, just because they did not know algebra or letters. In our part of the world, we had been used to associating illiteracy with the brains of children under the age of six. If an adult were illiterate, it was because something was wrong with his wits. But this was not the case with our friends in Ouia. If anything, *we* often felt absentminded and stupid, because they, at a mere glance, could see a solution to a practical problem that had dumbfounded us. We enjoyed their company.

After all, we began to realize, there is more to learn in this fabulous world than any one person can cope with, so it is for each of us to make the wisest or most advantageous pick of what to know and what to ignore. An astronomer knows the distances to the stars, and a botanist the number of petals on a flower. But neither dismisses the other as an ignoramus because his knowledge is confined to a different field.

We began to accept our two neighbors as specialists. Specialists in how to live and adapt in the best ways to the environment of the Ouia Valley. They did not know the size of either molecules or stars, nor the distance to the moon. But they knew the size of the booby eggs and ripe husk tomatoes, and they could tell me how far it was to where the nearest mountain pineapples grew. We felt ashamed to admit to ourselves that, with all our training and despite white man's inherited conviction that he had been born to remodel the earth, we from the world of letters ought to tread far more carefully outside our own circles. There is still so much to learn about life around us that is not yet spelled out in letters. How can we be sure that we do not destroy it before we understand it?

Tei Tetua had no shoes. He did not even own a pair to put on when he would have to crawl into his coffin. His only wardrobe was one strip of loincloth, which he occasionally wore like a diaper, but he was clean and he behaved as if he owned the world. The old Greek philosopher Diogenes, looking in vain with lanterns for honest human beings in the crowded market-place, would have found one in the Ouia Valley. Diogenes would also have given him a place in the sun next to himself in the barrel, because no king, no merchant, no teacher from any nation could have improved upon Tei Tetua's existence.

He accepted with gentle courtesy the little gifts I handed him on our arrival, but we never saw him use them. He took more pride and pleasure in anything he could pass on to us in return. He sometimes smoked his own homemade cigars, and the only one of our presents we eventually saw him use was a pipe. But he refused to accept any supply of tobacco, as tobacco plants grew wild next to his house.

The dead island of Motane was far away, and so was the unfortunate community of Omoa, but still close enough for us to realize how lucky we were to have ended up in Tei Tetua's world. We would probably have taken things more for granted had we landed here directly upon our arrival from Norway. This was Polynesia as we had expected to find it. It was a surviving fraction of a world that the white man likes to dream about, and at the same time wants to improve.

As I began to roam the valley with Tei to harvest our daily bread, it was easy to see that, because there was more sun and less rain, there was not the same tropical luxuriance in Ouia as there was in Omoa. Nevertheless, here too the landscape was a successful blend of wilderness and abandoned gardens enclosed by steep mountain walls. There was a superb variety of fruit, among which only the *fei* was lacking. There were breadfruit and bananas. There were mango, papaya, guava, husk tomato, and mountain pineapple. Taro fields, lemon groves, and large orange trees loaded with a kind of juicy, yellowfish fruit we never tired of. Tei also knew a number of edible leaves, bulbs, and some delicious roots and tubers. A few steps away from the houses, we could catch prawns in the river, collect the eggs of sea birds and chickens among the rocks, and we could snare the trotters of lazy boars.

We could fish from the cliffs, but the old man himself was not a keen fisherman. As so often in Polynesia, he left the gathering of coastal seafood to the women. When the ocean was not too ferocious, Momo and Liv would venture along the rocky promontory that projected from the southern corner of the bay. Here the lava had once hardened into strange formations: grottoes, tunnels, ridges, and depressions. At high tide, the spray from the sea would replenish the puddles and pools with clean, clear salt water. Myriad forms of life had assembled through the ages. Each pool was a natural aquarium, as colorful as a painter's palette, but the lava background was rusty-red and black. There was no coral. Among the myriad of little fish, octopuses, crustaceans, and mollusks, Momo knew what was worth catching.

To our surprise, Momo never worked in the kitchen. It was Tei Tetua himself who was the cook. They took turns in carrying the food to us. Tei and Momo ate by themselves in their own kitchen. We recognized the familiar smell of sour black *poi-poi* as Tei dug it out of his storage pit in the earth. Tei had to eat *poi-poi* with every dish he prepared. Like so many of his compatriots, he insisted that he could not digest a meal without a fair helping of this fermented dough.

Liv had no objection to Tei Tetua's table manners, for he washed his hands before he ate and we ate with our fingers as he did. But she said that whenever he chewed on a piece of sugarcane or gnawed at a bone, he would squat with his head tilted to one side just as if he were gnawing a human leg bone. The accusation was unfair and cruel, but once it had been pronounced I could not see our friend eat without the same silly associations.

Tei was not only a well-meaning host, but also a gourmet, a gourmand, and a truly capable cook. If we had experienced a scarcity of food in the Omoa Valley, that was far from the case here. Every day, in the morning, at midday, and at night, he or Momo came climbing up the ladder of our pole cabin with the most appetizing dishes. We even grew really fond of *poi-poi*, so long as it had been pounded together with enough fresh breadfruit and water. Tei's specialties were pork baked in banana leaves, soft-shell crabs boiled in coconut sauce, and raw fish soaked à la Tei Tetua. He selected his fish carefully, cut it into small cubes, soaked it overnight in undiluted lime juice, and served it in a mixture of seawater and coconut milk. Not the slightest taste of raw fish remained.

But no matter what Tei served, we always ate pork as a second course. At all meals. Juicy hot chunks of jungle boar baked in big leaves between red-hot stones. We received an abundance of food at every meal and we succeeded only in making a clearly visible inroad into the prodigious helpings and then prepared to take the rest back. But this was not accepted by the old man. We had to save the leftovers for our next meal. At the next meal, however, Tei would sometimes show up with a whole baked chicken, together with taro, breadfruit, and still more roast pig.

'The old man is intentionally fattening us up,' said Liv one morning, when she had trouble tying on her *pareu*, and had gone to a calm pool to take a look at her face. I was never quite sure whether she had only been joking, for the simple reason that she actually did go on a diet from that moment. For a fortnight, she ate nothing but large piles of oranges and pineapples, snatching only an odd banana from the cluster hanging from our ceiling.

As darkness descended over Ouia, we dumped food from the top of our ladder. We could never persuade the old man to take it back. Bristly boars from the forest crowded around our pole cabin every night and grunted, smacked their lips, and shrieked so much that we were worried that they might awaken the old man across the river. When the fattest of these nightly visitors scratched themselves against the wooden framework below, the whole structure would still sway like a crow's nest.

But the bark-lashed pole cabin was solid. Even when a blast from a tropical storm created unrest among the crowns of the jungle roof and flexed the tallest palms like archers' bows, we lay as safe as babies in a rocking cradle. Only if the storm were strong enough to hurl cascades of sea spray and rain against us horizontally did we have to get up and hang our pandanus floor mat against the seaward wall. The open side of the cabin, where the ladder came up, was on the lee side, where rain never entered. There was nothing, however, to prevent the moon from finding us, right in the innermost corner, whenever it rose to hang as the sun's mirror over the black silhouettes of the palms. The moonlight could always come in, yet not one mosquito flew through the open wall.

On many nights, the moon would rise to light an empty cabin, while the four residents of the Ouia Valley sat together around a campfire down by the beach. We were sitting in the orchestra, with a mighty stage in front of us. The stage was so big that people watched the same show from the Sahara, Greenland, the Amazon, and Fatu-Hiva. It was just about the only show that has united people from all over the world since time immemorial. When it started, Arabs, Eskimos, Indians and South Sea islanders were for a while on the same flying carpet, sharing a common universe, carried far away from the day's trivialities. It is not so strange that for many ancient people the moon was the goddess of love and the soothing mother of the universe, while the sun was the alert and industrious father. Only modern man has traded away the night sky in an attempt to obtain continuous day. He turns night into day in less than a second and puts on a million city lights until he sees nothing but his own world.

As if he did not want to dim the moon- or starlit surroundings with his own light, Tei always made a very small fire. Just big enough for us all to get close to one another and enjoy precisely the right amount of light and warmth.

Nothing could beat the nights when the moon was full and hung patiently, scattering gold and silver over the Pacific Ocean before us, twinkling behind us too, in the glistening palm crowns moving lazily against the stars. When the moon was full enough, it lit up the whole forest. Giant banana leaves and strange trees stood there in bizarre, sparkling nightgowns, crowding inland as far as the black silhouettes of the tooth-edged mountain wall that shut us out from the rest of the world. Except for the wind, the surf, and our own voices, there was no sound but the rare, occasional bleating of wild goats high above us, sometimes the grunting of boars, and, of course, the melodious chuckling of the river flowing nearby.

Tei Tetua possessed one foreign invention: a flint and steel that some early European voyager must have brought to his ancestors. He rubbed fire between two hibiscus sticks only to show his incredible dexterity in this old Polynesian art, but it was faster and easier to knock the flint and steel together and catch the spark on some tinder.

One evening, as the campfire was slowly dying down, Tei sat gazing into the embers, then he began to sway slowly but rhythmically, and he started singing in his coarse old-man's voice, a song that at first gave us gooseflesh, because it seemed to come from another world. But soon we were fascinated. There was no real melody and not many notes to his tune; it was almost like a liturgical chant set to music. Tei was singing about the world's creation:

> *'Tiki, the god of man,*
> *who lives in the sky,*
> *made the earth.*
> *Then he made the waters.*
> *Then he made the fishes.*
> *Then he made the birds.*
> *Then he made the fruits.*
> *Then he made the puaa* [the pig].
> *Then he made people:*
> *one man, his name was Atea;*
> *and one woman, her name was Atanoa.'*

Tei interrupted his song with the remark that these two made the rest of mankind by themselves. Then he continued his chant with a seemingly endless genealogical list of kings and queens who descended from Atea and Atanoa, down to the generation of Uta, Tei's father.

'Tei,' I asked, 'do you believe in Tiki?'

'*E*,' said Tei. 'Yes. I am a Catholic like everybody today, but I believe in Tiki.'

The old man grabbed a stone and showed it to me. 'What do you call this?'

'*Sten*,' I said in Norwegian.

'We call it *kaha*,' he explained. 'And this?' He pointed at the fire.

'*Ild*,' I said.

'We call it *ahi*,' he answered, and then he wanted to know the Norwegian name for the god who created man.

He has many names in my language, I told him, and gave him the Norwegian terms for 'God,' 'Our Lord,' 'The Creator,' 'the Almighty' and 'Jehovah.'

'My people call him Tiki,' was Tei's prompt reply. He insisted that his own tribe had understood that the missionaries were referring to Tiki when they came and told them about their own god.

I tried carefully to suggest that his ancestors had other gods beside Tiki: Tane and Atea were Marquesan gods also.

Important kings became like gods after death, but Tiki was the only creator, Tei explained. Tiki created Atea and all common people on the island descended from him. Tiki also created Tane, and he was white with fair hair, and the *hao'e*, white people like us, descended from him.

Tei took up his bamboo nose-flute and began playing a melodious tune with one nostril. He did not want to discuss religion any further. He was a Catholic, and Tiki was God. Tiki had led Tei Tetua's ancestors across the water to these islands.

'From where?' I asked, and was curious to hear what the old man would reply.

'From *Te-Fiti*, the East,' answered the old man and nodded toward that part of the horizon where the sun rose. In the direction where there was no land except America.

I was amazed, but not surprised. Henry Lie had toldme this was the belief also among old people on Hivaoa. The American ethnologist E.S.C. Handy had also obtained the same information. He even recorded a tradition on Hivaoa about a return voyage to the fatherland Te-Fiti. A party of men, women, and children had embarked from Atuona bay in the Kaahua, a vessel of extraordinary size, to visit their ancestral land. They sailed east until they finally came to Te-Fiti, and there some of the voyagers remained while the others returned to Hivaoa. Handy wrote that he became so puzzled about the claim that the fatherland of the people in the Marquesas was to the east, that he had to ask twice, and his informant insisted that this land was 'toward the rising sun,' *i te tihena oumati*. A generation before Handy, the German ethnologist von den Steinen had been equally surprised to learn that the Marquesas islanders spoke of a large legendary land called *Fiti-Nui*, the 'Great East.'

Now it was my turn to get the same information on Fatu-Hiva, where none of the others had been. I looked across the moonlit sea in the direction the old islander pointed. America was there. Nothing else but open water. I was sitting on the one island nearest to America, the Polynesian outpost first sighted by the Europeans when, following Inca sailing directions, they came from Peru.

I really started to wonder why we from the outside world ignored native tradition and overlooked the lessons of our own maritime history in the Pacific.

When Tei Tetua told me that Tiki had led his ancestors over the ocean from the east, I was ignorant as to how well this fitted Incan traditions. Not until I went back to the library shelves of the outside world did I find that Incan history began with a legendary king who had ruled in Tiahuanaco. The Incas came to power when he descended to the coast and left with his fair-skinned followers, going out into the Pacific. The Incas called him Viracocha, the same name they gave to the Spaniards when they arrived. But his original name in Tiahuanaco was Tici or Ticci.

It was he and his pre-Inca people who had carved the great stone men in the Andes. He was celebrated in every part of the vast Incan empire as a personification of the creator, but also as the human progenitor of the royal Incan line.

The past and the present seemed to melt together as Liv and I were sitting beside Tei Tetua and Momo under the starlit sky by the boulder beach. As the old man passed on to us what he had learned from his father and his tribe, his words were accompanied by the perpetual drone of the surf pulling against the rattling stones. I began to understand that I was listening to fragments of Polynesian history, although to the outside world this would all sound like fairy tales concocted by Tei's illiterate ancestors. In Polynesia, as in South America, great kings were gods, whether dead or alive. Perhaps Tiki had really been some important personage who had come with the winds from Fiti-Nui, the Great East, as these people kept telling us on Hivaoa and Fatu-Hiva.

As the days passed in the company of old Tei, I began to look at Fatu-Hiva's history with the eyes of the local people. Of course the Mendaña expedition had not discovered Polynesia. It was Tiki, whoever he was, who had discovered Polynesia. Like Mendaña, he had also come from the east, from Fiti-Nui, but so long before Mendaña that the descendants of his descendants were here to greet the Europeans when they came.

We may forever doubt that the Inca historians knew what they were talking about when they taught their growing generations that Ticci, the founder of the empire, sailed west into the Pacific. To us, this is the tale of illiterates, like the tale of Tiki who came to the Marquesas from the east. To us, Mendaña was the first to venture into the ocean from Peru. And to us, he was the discoverer of Polynesia. How there could be human beings on Fatu-Hiva before they were discovered by us is more of a mystery to us than to the people we discovered.

Tei used a peculiar term for the early Europeans. Those we refer to as the 'discoverers' of Polynesia, he referred to as the 'double-men.' I felt rather ashamed but thought it was a fitting name for us Europeans. First we had come to them with priests and told them *not* to kill. Then we came back with army officers to show them *how* to kill. We teach them that the Bible tells us not to accumulate earthly wealth. But we ourselves want insurance and money in the bank. We admit that God made mankind naked, but when we see naked people we dress them up. We arm for the sake of peace, and we lie in the interest of truth. Double-men, that is just what we are, I thought.

But I got a surprise when Tei explained the Polynesian reason for the name.

When his ancestors had seen the first Europeans, he said, they seemed to have two heads, two bodies, four arms, and four legs. The islanders had never seen tight-fitting clothing such as that worn by the first foreign visitors. When they took off a helmet, they had another head underneath; when they loosened their armor or unbuttoned their clothes, another body appeared; and when they pulled off their boots, an extra pair of feet turned up inside. This had caused great astonishment among the islanders.

But the double-men had brought coughs and fever. Nobody died from illness before they came, Tei Tetua insisted. People used to grow so old that they would sit in one spot like dried calabash rinds and let others feed them. If someone died young, it was because he fell from a tree, was caught by a shark, or was hit on the head by a club and eaten by the enemy.

'Eaten?'

Liv shook her head in horror at such a statement.

'Don't you make war in your countries?' Tei asked, his expression seeming to say, 'Now come on, tell the truth!'

I had to admit that when we had left Europe a fierce civil war was raging in Spain.

'And what do you do with those you kill?' Tei wanted to know.

'We bury them.'

'Bury them!' Tei was amazed and truly disgusted at such barbaric waste. Imagine killing people only to bury their flesh in the ground. Did nobody come and dig it up again when it was matured?

Was Tei being sarcastic or did he mean it? He appeared to be serious. He seemed to look at us as we look at a Hindu who leaves a sacred cow to bugs and dogs once it dies.

Tei told us about his father, Uta, who was the greatest warrior of the Ouia Valley. He rarely ate any meat except for human flesh. Unlike his friends, he wanted it old and tender before he headed for the burial platform to fill his *poe* or calabash bowl. He would eat fermented *poi-poi* together with this foul meat. Once when a member of his own tribe had been killed in an accident, the widow came to present Uta with a pig. She wanted to safeguard her husband, and thought that Uta would make do with the pig until her husband was gone. But Uta ate the pig first, and after the pig was gone, he ate the husband. Tei's mother was furious at her husband, Uta, and begged him to eat fish and decent food that did not stink. And Uta was kind, said Tei. For many days, he did not touch foul flesh. But then he got very ill and skinny and had to return to his customary diet.

Liv was horrified, and Momo sat with huge brown eyes and a half-open mouth, looking, like us, at the peaceful old man who sat there calmly recounting these events as if he were telling us about a meal at a faculty cafeteria. Tei himself had only participated in a cannibal ceremony once, here in Ouia when he was young. Human flesh was sweetish, like *kumaa,* or sweet potato. The victim was usually baked, like the pork he prepared for us. That is, rolled up in banana leaves between hot stones in an earth oven. Some people ate human flesh from hunger, as at that time there were too many people for all of them to have enough food. But usually human flesh was eaten as a religious ceremony, and as a sort of revenge.

The choicest piece was supposed to be the forearm of a young woman. 'A white woman,' Tei added, looking at Liv with a grin. This piece of information was obviously meant as a joke for her benefit, but I doubt if it was very much appreciated by either of the two women. I threw another stick on the embers to get a bit more light. Tei was undoubtedly the finest man on the island, but sitting there under the stars hearing cannibal stories told in the first person had a strange effect on me.

Whether Spaniard, Polynesian, or Viking, man has always been a strange mixture of saint and Satan. One moment we can be so pious that we do not want to cut a lock of red hair from another, and the next moment we murder, bury each other in the ground, or roast each other like pigs.

Tei Tetua had a deep round dent in his forehead. I asked if somebody had hit him with a club, but he said he had been operated on by a *taoa* when seriously hurt by a falling stone. I knew that the *taoa*, or Marquesan medicineman, formerly had a reputation for his considerable insight in both basic psychology and advanced surgery. The Marquesas was one of the few regions of the Pacific, apart from ancient Peru, where trephination was formerly practiced.

Today it is different, said Tei; today a mere cut or scratch will cause a *fe-fe* infection. But *Taoa* Teke had cut open and cured Tei's head, and he had even seen him cut open the leg of a man who had broken his shinbone; he had spliced the fracture with a piece of hardwood. *Taoa* Teke had lived in Ouia but was buried in Hanahepu under a small stone Tiki that marked his tomb.

Tei used the term *taoa* as we would have used the term 'doctor,' and with good reason, for Teke performed trephination in Ouia in the second half of the nineteenth century, when it was still an art regarded as extremely difficult in our own part of the world.

Tei had seen him cure the fractured skull of a young man who had fallen from a palm. After appropriate dancing and incantations with a bowl of steaming water, he began to wash the wound, and removed all hair from the injured part of the head. Then he cut a cross-shaped incision in the skin over the fracture and exposed the skull. Splinters were removed. The edges of the fracture were polished until the opening was smooth and round. A thin disk of polished coconut shell carved to fit was put over the opening, and the four flaps of the skin were bent back over it again. The skin healed, and the man survived with a scar like a cross on his head. Tei had known him. He was a bit odd, he admitted.

*

One day Tei wanted me to join him on a climb for *faa-hoka*. These were the wild pineapples, of a different kind from those we used to eat on the other side of the island. The Ouia ones were smaller, but superior in fragrance and flavor. To find them we had to leave the valley and climb the steep and treeless slopes of Natahu, a lofty mountain pyramid towering above the sea on the south side of Ouia. It rose like a green spire on the long island roof of Tauaouoho.

The trade wind tore at our hair and our loincloths as we crept up the stony terrain to an altitude of almost a thousand meters over the sea. The ascent offered a remarkable view over the expanses of the Pacific. Up here was what seemed to be an abandoned pineapple plantation growing haphazardly among the rocks. Thirsty after our hot climb up the sun-baked hillside, we lay down and reveled in consuming as much as our stomachs would accept of this incredibly juicy product of the arid ground. With burning lips we crawled around and filled our plaited bags, then lay for hours half dozing and gazing out at the ocean. The island seemed to be sailing eastward against the endless stream of breaking wave crests down below, and against the endless multitude of trade-wind clouds that moved above us like millions of sheep trying to jump the barriers of the Fatu-Hiva mountain ridge. Some failed in the jump and huddled together as in a slaughterhouse, sending their tears as rain down the slopes on the other side. Our side remained dry. The jungle grew on the Omoa side.

Half-awake, I abandoned the feeling that Fatu-Hiva was drifting. I lay on a rock in the middle of a vast river. As I sat up and looked at the waves and the clouds I suddenly and excitedly realized that the largest river of South America was not the Amazon. It was the Humboldt Current. Both start in Peru, but in opposite directions. The muddy brown rainwater of the Amazon shows up clearly as it flows east between the green jungle banks of Brazil. But the much wider and equally fast Humboldt Current bends away from the coast of Peru and flows westward to embrace in its sweep the Polynesian islands. Only the temperature and its richer plankton content distinguishes this moving mass of once-Antarctic seawater from the more stagnant banks of blue ocean through which it flows.

No wonder that east was 'up' and west was 'down' to the Polynesians. They lived downriver from Peru. Of course that was why the botanists had found that most of the flora of the Marquesas had come from South America. I knew that even the *pavahina* grass that covered the slope we were lying on had come from South America. Nature had possibly arranged for the downriver transport. But not for that of the *faa-hoka* pineapple.

It dawned upon me that the pineapples I had come with Tei to pick had a direct bearing on the origin of the people who had begun to absorb my main interest. The pineapple was a South American plant. It could not have spread across an ocean without human aid.

When I prepared for this voyage to Polynesia, most of my training had been in biology. I had plowed through three thick volumes by the American botanist F. B. M. Brown on the flora of the Marquesas. He had shown that there were two kinds of pineapple here: the larger commercial type of *ananas* had been brought to the Marquesas by missionaries from Hawaii in the last century. The other, a smaller form with six local varieties, had grown in a semiwild state throughout the Marquesas before the Europeans arrived. For purely botanical reasons, Brown had concluded that the pre-European introduction of pineapple to the Marquesas was a genetic proof of contact between South America and this island group before the Europeans came.

I had been puzzled when reading Brown's statements back home, but I had been more interested in the biological problems and in my own future survival on these islands than in the sailing routes of ancient man. Now, lying on the windswept slope and enjoying the presence of wild pineapples on Fatu-Hiva, things looked different.

'Tei,' I asked, 'did the double-men plant the *faa-hoka* up here?'

Tei looked at me as if I had asked a silly question. '*Aoe*,' he said. 'No. The double-men never climbed here to plant anything.'

From this moment I began to think that botany was a field that could give conclusive evidence to the anthropologist trying to trace migration routes across an ocean from which all traces of voyage had vanished. When I came to think of it, I had taken note of other edible roots and plants that Brown had shown to be of aboriginal introduction from South America. The papaya was an example. This, too, was a strictly tropical American plant, important to the coastal people in ancient Peru. Two varieties of papaya grew in the Marquesas. The natives called the larger and more tasty type *vi oahu*, and said that it had been brought from Oahu in the Hawaiian group by the missionaries. Their own smaller type they called *vi enata* and recognized it as one of the ancient food plants their own ancestors had brought from the fatherland. *Enata* means 'people,' and was the word the Marquesas islanders used for themselves and their own kind. Brown had concluded plainly that this type of papaya represented another introduction from South America by aboriginal man.

And he had pointed out other useful plants in the same category. The sweet potato, for instance. The delicious root we had eaten so much of in Omoa. When the Europeans arrived, they found the sweet potato, one of the oldest and most important cultured plants in ancient Peru, to be the principal food plant from Easter Island to New Zealand. Even the anthropologists, after much vivid controversy, had been forced to admit that the American sweet potato, with its aboriginal South American name, *kumara*, had been brought from Peru to the widely separated islands of Polynesia in pre-European times.

The calabas too, the edible 'bottle-gourd' whose fire-dried rind was used for water containers on every Polynesian island, had caused angry dispute among botanists and anthropologists. The anthropologists had argued that it must have been carried from Polynesia to Peru by European ships in the early years of discovery. The archaeologists had settled the dispute. Dried rinds of calabas fruit used as in Polynesia for containers and as floats for fishnets, were excavated from tombs on the desert coast of Peru that dated back to the third millennium before Christ. Wood-boring worms and sharks would have ended the life of a calabas long before the Humboldt Current could deliver it from Peru to the Polynesians, so man must have helped it across the Pacific before the Europeans came.

For over a hundred years scientists had also disputed the origin of the coconut palm, so important on all the islands straight across the Pacific. Botanists had found that the only place where it grew wild was in Colombia, and all its related genera, including about three hundred species, were found to be American. From there it had entered the open Pacific – or even spread to Cuba, where Columbus found it. The Polynesians had dispersed it from island to island in their part of the ocean, but only the Marquesas islanders had a tradition about whence it came. The early voyager Captain Porter was told that the first coconuts were brought to them, not from any other island in Polynesia, but from a distant land to the east. And not by canoe, but by a vessel that Porter thought was carved from stone, as the islanders used the term *pae-pae*, which is their word for a stone platform. But it is also their word for raft.

As Tei and I slid side by side down the slippery *pavahina* grass with our loads of pineapple, I had a double reason to stop at an area rich in shiny red husk tomatoes, the size of big berries growing on the ground. These were Liv's favorite. The flavor and appearance was that of a minute tomato, of which it was, in fact, a wild South American ancestor. Here was one more of the plants Brown had used in his argument about aboriginal voyages from Peru, as it was found from Easter Island and the Marquesas to Hawaii when the Europeans entered the Pacific. As I picked a few and put them in with my pineapples, I could never have guessed that the next time I was to find them growing wild would be in the 1990s, on the site of my new home at the foot of the Tucume pyramids in Peru.

Yet I did know, as I handed my heavy bag to Liv, that its juicy contents were going to have an important impact on my life. My interest in transoceanic migrations had come full circle, from animals to man. The problems of the diffusion of humans appeared to be linked to the spread of cultivated plants. I was back in my own field, genetics and biology. Man can create the same kind of tool on either side of an ocean. But a pineapple he has to bring with him.

Brown had clearly not been able to impress his anthropological colleagues at the Bishop Museum in Hawaii with his botanical arguments. It was most unlikely that any of them had ever read his three volumes on Marquesan plants. To them the Pacific ended with the Marquesas and Easter Island. To anthropologists there was an imaginable abyss between Polynesia and South America, impassable for aboriginal watercraft. Brown, as a botanist, had to accept their views on human sailing routes. But he clung persistently to his own conclusions:

'Although it appears that the main stream of Polynesian immigration came from the west, just the opposite direction from which the indigenous flora came, undoubtedly some intercourse may have occurred between the natives of the American continent and those of the Marquesas.'

His argument impressed me more than the only counter-argument the anthropologists had: the unproven dogma that the ancient Peruvians had neither the courage nor the watercraft to depart from their own continental shores.

From then on I began to suspect that the Polynesian riddle could never be solved by a specialist who put his head deep into one narrow hole. We do need the analyst who can dig deeply into his own field. But we also need another type of specialist: the academically trained generalist able to make a synthesis by combining available information from all fields of science. To reconstruct a complete picture of the unrecorded history of Polynesia, we need teamwork – scholars working as scientific detectives, ignoring no footmarks or fingerprints.

I felt that universities ought to open a new kind of faculty: one which could do horizontal research in addition to the existing vertical kind, crossing disciplines as I felt earlier man had crossed 'impassable' oceans.

'Liv,' I said as we crept to bed, 'do you remember how Hivaoa's big stone statues resembled those of South America?' I could not refrain from bringing this theme up again. Liv, falling asleep, merely grunted. Only the rumbling surf seemed to reply in approval.

I could not sleep. I felt as if time no longer existed, as if Tiki and his seafarers were standing in the bay with full sails. Red- and black-haired men and women jumped ashore on the boulder barricade. They unloaded baskets full of roots and fruit to be planted.

I felt for my own pile of pineapples and tomatoes. They were there. They were real.

I turned over on to my side and fell asleep.

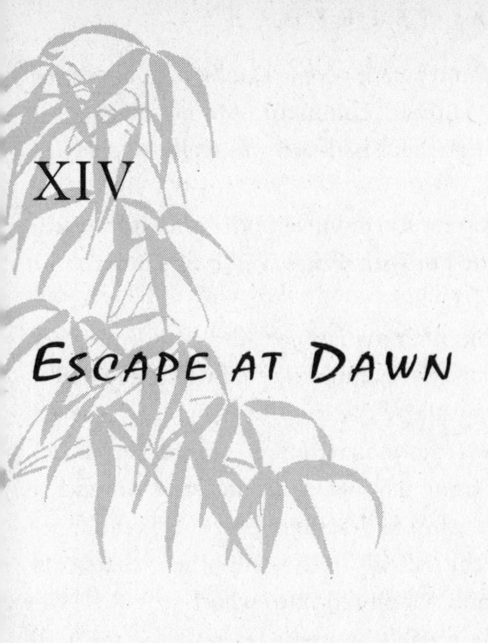

XIV

Escape at Dawn

HAPPY DAYS MADE TIME BLUR in Ouia. The tropical sun rose straight from the sea and sank behind the mountains as the pale moon followed across the night sky. But there was no tomorrow nor any yesterday. Everything was today, no matter how often we saw the starry heavens light up and fade again at the colorful approach of another sunrise.

Our former life was like a remote dream. A distant yesterday. Civilization was incredibly far away. When we thought of it, it gave us a strange feeling, as if we were in a science-fiction novel. When we tried to describe our own world, our former life, to Tei and Momo, they never seemed to quite believe our words. And it was difficult for us, too.

Back from the forest one afternoon, we had a bath in the river and lay on the grass gazing at black frigate birds with scissor-shaped tails cutting through the clear air.

'Tei,' I said, 'in my country even people can fly high above the treetops.' Tei groaned without comment. Momo laughed. I started to wonder whether what I had said was really true. Had I merely dreamed it?

'It is true, Tei,' I insisted, to convince both him and myself. 'We just crawl into a little hut with wings, and it lifts into the air with us inside.'

I tried to explain to him that my father had dared my mother and me to climb into such a flying building that was so big that it lifted four people up to fly above the roofs in the village where I was born. Its wings were motionless, but it moved because of something crooked in front that went around and around. I waved my arm in a futile effort to describe a propeller, simultaneously realizing that on this island there was nothing that went around, neither a windmill nor any form of wheel.

Momo continued to be terribly amused; Tei looked at me with a twinkle in his eye and wound his arm understandingly around and around. Of course, I realized, he neither understood nor believed.

I was going to add that one man had already flown alone all the way from America to Europe, but realized that neither America nor Europe meant a thing to Tei and Momo.

I climbed up into our cabin to fetch some crumpled pages of an old magazine given to us in Hivana. It had no pictures of airplanes. But there was a good picture of the New York skyline.

'See the size of these houses,' I explained and pointed at the skyscrapers of Manhattan.

Tei and Momo were unimpressed. Tei finally grabbed the paper, turned it upside down, viewed it from all directions. No reaction, until he turned some pages and saw the picture of a man and woman in front of their two-story suburban home. Then Tei and Momo got equally excited. A house on top of another! Never had they heard of houses that big. They never tired of admiring the two-story residence. The Manhattan skyscrapers, without any human beings to indicate the scale, meant nothing to them. As I took a second look at the picture myself, I got that eerie feeling once more. Was this true? Had buildings like that already been built on the same planet we were on? I remembered our own world as if it were a vision into the future.

Our lives continued happily in Ouia. It was difficult for us to explain to Tei why we in the outside world worked from morning to night to buy a hundred things, and yet we felt happiness like sunshine inside ourselves, day and night, by living as he did. Every day we roamed among rocks and foliage, enjoying life as we provided for ourselves and our women. But I did not like to join Tei when he went with his dogs and his ropes to catch a hog. I realized I was a hypocrite not to join him, for I enjoyed eating his baked boar. I preferred relaxing on a fallen moss-covered tree or the smooth stone platform of an abandoned *paepae*. No matter how much we struggled with our bodies, I felt comfortable.

Here, I thought, we keep our bodies fit so we actually enjoy the sweating and panting that make us suffer at home, and we use our rested minds to enjoy all we see, hear, smell, and feel. This is a fine way of providing a livelihood. Tei and I go hunting and fishing, we pick fruit and berries, we stroll in the woods, climb the rocks, we swim; all for a living. What we do for work is what others do on their vacations. They sit all day at a desk or stand at hot conveyor belts in a factory eleven months out of twelve to earn money for cars and big houses. Then, with the money they've saved, they rush away from their big houses for a few weeks' vacation in a little cabin or tent. To look for a place in the sun. To hunt, to fish, to pick berries, to stroll in the woods, climb the hills or swim. Primitive man's work has become modern man's leisure. Even spring water, clear air, and sunshine is luxury to modern man. He locks himself up indoors with artificial light so that he can earn enough money to pay his electricity bill and a few weeks outside in the sun.

I did not tell this to Tei. How could he understand? How

could I confess to him that we prefer to sit all day at a desk and do a job that involves pressing buttons, only to get up after work to start physical exercise, jumping rope in a frenzy or lifting heavy iron rods up and down above our heads? It would seem odd to him if I told him that we put motors on our bicycles because we are too lazy to pedal, and other motors on our boats because we dislike to row. Then we buy costly physical fitness gear to put in our basements, where we sit in one spot and pedal on a bicycle without wheels, or row a boat without bottom.

One day, when I was sitting pondering man's many bizarre ways of harvesting his daily bread, I heard Tei shouting somewhere far away. His dogs were barking ceaselessly. The old man needed *help*. I ran as fast as my lungs permitted across the valley to the mountainside, from where the angry, incessant barking was coming. I found Tei waving, unhurt at the foot of the cliff, his two dogs dancing ferociously on their hind legs, trying to get up on a rock shelf. A beautiful shaggy goat, white as snow, stood on it with head and horns lowered, ready for brave self-defense. When I joined him, Tei sneaked up from behind and grabbed the goat by its back feet, holding it firm until I succeeded in getting a good grip on the horns. The goat was ours.

It was a hard fight to keep the dogs away and get our struggling booty down to the coast, where Liv and Momo helped us to tether the pretty animal to a pole beneath our cabin.

'Now we can get milk!' Liv exclaimed enthusiastically. Momo bent down, then shook her head. There was no milk to be had from a billygoat. Liv offered the goat a banana, and it ate. Before evening, the wild creature was relaxed and unafraid, its belly taut from fruit and taro leaves. We had our first domesticated animal friend, and called him Maita, meaning 'white'.

Weeks passed. Weeks so full and rich that each month felt like a satisfying life span. There was no shop, no market, no middleman, no expenses. It took time and energy to search for and harvest our daily needs, but we collected animals and artifacts on the same trips, and still there was time for rest and entertainment. The land snails and insects were different in many ways from the fauna on the other side of the high mountains, but the tools and house platforms left behind by the former people were the same: stone adzes and pounders, grooved net sinkers and octopus lures, shell scrapers and worm-eaten woodcarvings.

A common implement was a perforated stone disk, the only artifact Tei could not identify. We had also found a great number of these in Omoa, and nobody could tell us what they had been used for. They resembled a certain type of South American spinning wheel, but the Polynesians knew neither the art of spinning nor weaving when the Europeans found them. Nevertheless, it was a striking fact that cotton grew wild on several Polynesian islands, notably in the Marquesas group, when the Europeans arrived. In Tahiti the missionaries, surprised to find spinnable cotton, tried to encourage the Polynesians to start to harvest it for export. But the islanders did not bother. It was enough for them in their warm climate to dress in loincloths or make themselves ponchos of beaten bark.

Could it be that the people who brought the pineapple also brought the cotton and used the stone disks for spinning until the *tapa*-beating Polynesians reached the Marquesas? At any rate, the cotton could only have reached these islands from South America, for neither wild nor domesticated cotton grew on the opposite side of the Polynesian territory. Not until a decade later did a group of cotton geneticists seriously study the anatomy of the world's cotton species. I had long since gone back to Europe when Hutchinson, Silow, and Stephans published their discovery that the pre-European cotton of the Marquesas was derived directly from the cultivated cotton fields of pre-Columbian America. All other cotton species in the world has thirteen chromosomes. But the pre-Columbian civilizations in Mexico and Peru had managed to hybridize a long-linted, spinnable kind with twenty-six chromosomes, and it was this cultivated American cotton that had reached Polynesia and run wild even in the lonely Ouia Valley before the Europeans came.

But I was ignorant of this discovery. I picked no cotton from the bushes, nor did I see Tei use this plant or find an intelligent use for the perforated disks.

Back in our pole-cabin late one afternoon, I lay musing about the origins of the early seafarers who had lived in this valley before us. I began to see ever more problems in ascribing their origins solely to Indonesia, a distant part of the world with a completely different culture. I knew that physically the Indonesians and Malays also differed from the Polynesians in almost every trait studied, according to the investigations of the physical anthropologist L. R. Sullivan of the Bishop Museum in Hawaii, the leading contemporary authority in this field of research.

I was abruptly interrupted in my musings and brought back to myself. I had heard something unusual. The distant barking of dogs. I heard them again, far away up the valley.

Tei was just wading over the shallow river to bring us a huge leaf topped with a steaming hot evening meal. The two dogs at his heels immediately stopped following him and threw back their heads to reply to their distant relatives with wild barking.

Somebody was coming, for we had never seen or heard wild dogs down in the Ouia Valley. I had thought I heard voices high up in the cliffs earlier in the afternoon, when I was in the valley cutting firewood. But then I had believed I'd imagined them.

The barking of the dogs rose to a deafening crescendo as a whole pack of white and spotted mongrels of the pointer type came out of the thickets and down through the open palm grove, a group of men, women, and children in their wake. They shouted, waved and saluted. They were our fine friends Ven and Tahia-pitiani from Omoa, together with another couple and a lot of children, among whom we recognized the little rascal Paho, Pakeekee's son. The calm valley became noisy with shouting and laughter.

Obviously, it was not the good wind of Ouia that had lured these unexpected guests over the mountains. We had hoped that the people on the other side had forgotten about us. Now we hoped that these visitors would never return to Omoa with tales about our pleasant life in Ouia, and we wholeheartedly supported Tei when he begged the visitors to stay: The wind was good in Ouia and there was plenty of pork. Breadfruit in the valley and *faa-hoka* up in the hills. Why go back to all the sick people in Omoa?

A new piglet had just come out of Tei's underground oven and the *poi-poi* was ready, so there was food for all. The newcomers needed no further persuasion. They entered the stone enclosure around Tei's residence with all their children and the pack of dogs, and when the piglet had been devoured to the last bone, we saw through the gaping bars of our hut that they all crawled into Tei's empty guest house.

The newcomers decided to remain. They built no new homes, but settled with Tei and ate with him. They were the finest colonists we could possibly have received from the other side. Friendly, energetic, healthy, and remarkably good-looking. Veo was the best hunter on the island, and although Tei's half-domesticated supply of boars was soon exhausted, there were still plenty in the valley. Veo's hunting equipment was his pack of dogs and a strong loop of hibiscus rope. The others were experts in climbing trees too difficult even for Tei to tackle, and they were incredibly skillful in supplying our joint household with fresh fish and other seafood. It was as if the good old days had returned to the valley. There were people up in the hills and along the seashore. Children were shouting and women were laughing. Tei was happy. We were all happy. We worked together and shared everything.

Some days, to our surprise, the sea became so calm that we could dive into the surf and swim. The clouds drifted sideways for a change. But only for a couple of days, and then they were back on their normal course.

Paho and the other children were masters in catching octopus. They ate them raw. Admittedly, they were a delicacy if cut into cubes that were soaked overnight in lemon juice, but to tease us they ate them alive. They chewed on the body of large octopuses while the creature twisted its tentacles around their necks. They were terribly amused if we shuddered at the sight. Momo immensely enjoyed discovering that Liv was ticklish when a straw was applied to the soles of her feet. She would shock Liv by taking a sharp flake of lava and cutting off a slice of her own foot, at the bottom, where the skin was thick as shoe leather.

In the evening around the campfire, we all joined Tei in singing the old melodies he taught us. Or we listened silently to tales about his childhood in this valley. There were schools on the island in those days. Regular schools, where the main activity was, under threat of punishment, to learn verbally and by rote the myths and traditions of former times. In those days, kings married their sisters and men were closer to the gods. There were plenty of turtles all along the coast, and people lived right to the top of the Natahu Peak. Today, up there, one could find nothing but some strange vertical shafts leading down to empty underground chambers. Times had changed on Fatu-Hiva. It seemed as if Tei had a silent wish that all of us together might now turn it back to what it was. The old man was more vigorous and active than ever.

On the other side of the island, people began to wonder why Veo and his company had not returned. A few weeks passed, and then more and more people began streaming over the mountains from Omoa down into the valley of Ouia. These newcomers included some of the worst troublemakers. They all accepted Tei's invitation and crawled into his two huts. Tei alone struggled in the kitchen. His latest guests did not even bother to help in providing food. Instead, they started to brew a kind of beer from our supply of oranges. They sat all day squatting along Tei's walls, or lay stretched out dozing, insisting on being fed by Tei and Momo while waiting for their brew to ferment. Alcoholic drinks of all sorts were unknown in Polynesia when the Europeans arrived. It was a noteworthy fact that the Asiatic custom of brewing palm wine and chewing betel with lime had never spread to aboriginal Polynesia. The Indonesians drank palm wine. And betel-chewing had spread as far into the Pacific as the islands of Melanesia. But there was a marked transition line between the West and the East Pacific. The Polynesians of the East Pacific, like the aboriginal people of America, were ignorant of alcohol. But they had another curious drinking habit in common. From Mexico to Peru and Chile the aboriginal tribes and nations produced a ceremonial drink known as *kasawa*, *chicha* or *kawau*, and in Polynesia as *kawa*. Throughout this American-Polynesian area this drink was produced by women, who masticated a root or other suitable vegetable matter, in Polynesia the root of the local *piper methysticum*, then spat the mash thus made into a bowl of warm water. When this soup was sufficiently fermented, the fibers were strained and sifted, and the result was a salivary ferment which was nonalcoholic but had a soothing effect on the consumers. As in ancient Peru, this brewing was an extremely important part of the local culture, and the brew was drunk ceremonially and in the honor of divine ancestors.

Unfortunately for us, the newcomers from Omoa were not preparing *kawa*, which would leave them quiet and drowsy. Over at Tei's place, men and women were just sitting waiting for their orange beer to ferment, so that they could have a real modern orgy.

We, on our side of the river, were worried. We knew that a drunken Polynesian could be up to anything, since he usually lost control of himself completely. Some of the most horrible slaughters, with people even eating other people's heads, had taken place in the Marquesas when these islanders first got to taste alcohol, a few decades earlier.

Then the drinking began in the crowded courtyard across the river. Tei still struggled alone in the kitchen, but was pulled away to join the party. He was made to drink with the rest, and even the small children were plied with the brew till they were completely intoxicated. Little Momo, too. A big islander of mixed blood, named Napoleon, was soon worse than the rest. He became completely insane whenever he tasted alcohol. He had whipped two wives to death when drunk. Now he was flirting with Hakaeva, a widow who had come across the mountains with the others. Père Victorin had been duly paid, before he left, to see to it that the husband got to heaven, so now she was back on the marriage market, with a flower behind her right ear.

That night several visitors tried to climb up our ladder, but they were too drunk to succeed and we pushed them down. The noise from across the river was incredible. If this company was to continue, we could not remain.

There was no sign that the party from Omoa had any intention of returning to their homes. On the contrary, those who were not too hopelessly drunk the next morning staggered into the woods to look for more oranges with which to start the next batch of brew. To our dismay, Tei Tetua came slouching across to our cabin and called me down from the ladder in a thick, hostile voice.

Red-eyed and reeling, all he could say when I climbed down to him was: '*Etoutemonieuatevasodiso*.'

He had to repeat it twice before I got the point. It meant 17.5 francs, the daily pay in Tahiti requested by Ioane when he and his friends had helped us put up the bamboo cabin. Obviously, Napoleon and his friends had sent old Tei over to demand daily wages.

'But, Tei,' I said surprised, 'you cannot use money.'

Tei hardly heard what I said. He just turned around and staggered across the river, mumbling once more, '*Etoutemonieuatevasodiso*,' as he left.

Liv was worried, and with good reason. As for me, I stopped troubling my head about where Tei Tetua's first ancestors had come from. I had no other question in my mind but where we ourselves could go. For we could not remain here.

A couple of days later, I was sitting on the ladder of our pole cabin, looking out to the open sea. Over at Tei's place, people were sleeping or sitting along the walls, waiting for the next lot of booze. A couple of women sat naked in the river and splashed water over their shoulders. Then I noticed a fine column of smoke far out in the ocean. A steamship! This was the first we had seen from Fatu-Hiva.

I had never before been able to understand why shipwrecked people sat ashore scanning the horizon for a ship that might pick them up, once they had been lucky enough to end up like Robinson Crusoe on a beautiful South Sea island. Now I did the same myself. I was sitting, bearded and long-haired, on the ladder of my Robinson Crusoe home among waving palms, following the column of smoke on the horizon, wishing we could get on board. The masts came slowly above the horizon, then the smokestack, and then the whole ship.

A steamship was approaching the island.

Liv was now at my side. The ship was heading obliquely toward Fatu-Hiva. Out there, on board that black hull, were people from our own world. They were surely lined up at the railings, watching the supremely lovely South Sea island, as we had done ourselves, when we came with the big ship to Tahiti. Seemingly a century ago.

Now, through their binoculars, they were undoubtedly staring at the pole cabin next to the beach. They surely took it for a native hut. For the steamer slid slowly by, far out at sea, probably bound for Tahiti.

Then we were left alone again, without anyone we could trust.

Next day, a young man took off from Ouia to return across the mountains. We asked him to take a letter to Pakeekee, but he refused. We offered to pay for this service, but he still declined.

Shortly after this, Liv woke up one night with a terrible sting in one thigh. She grabbed at the spot and told me the bed was full of insects. I realized at once what it must have been. She had not been stung by myriad little insects, but pinched by the poisonous jaws of a giant jointed centipede.

In the sparse moonlight we could not find the intruder among our palm leaves.

For the rest of that night we hardly got a wink of sleep for the racket that went on across the river. Napoleon's voice screamed and shouted above all the rest. We squeezed orange juice into the tiny double hole made by the pincerlike jaws. It both soothed and cured. Next day, Liv was only a bit stiff in the leg. As the first rays of the sun began to play through the sticks of the wall, we got up and checked every leaf in our bedding. Then we found the yellow centipede, rolled up alive like a tiny serpent beneath us. I severed its head with the machete. We went on searching and found one more, which I also killed, but still another big one wriggled down through the cracks in the floor and dropped away.

That day, we learned from a red-eyed and confused Momo that several big *poe* bowls with orange brew were ready for a much greater treat than the last one. Why did we not join in? I managed later to get hold of a young man who came alone to the river, and he confessed that he knew the trail back across the mountains. I bribed him to stay away from the party and follow us into the highlands that night before dawn. He consented after having been given advance payment.

A happy event had just taken place in the valley. A wild sow had just produced a sextet of piglets, and in the confusion, as the piglets ran in all directions when the dogs chased the mother, Momo had grabbed one and made a present of it to Liv. It was the cutest little creature, with laughing eyes and a happy grin on its long, thin snout. It had a coquettish curl to its minute tail, pink hoofs like a toy cow, soft fur like a red-haired boy with a crew cut, and was beautifully speckled with black dots. Liv was completely lost with tenderness, and tucked it safely away in the cabin. The piglet was adopted in the midst of all this tragedy and confusion, and given the name of Mai-mai, Momo's pet name for piglets.

The young man whom we had bribed kept his promise. Another wild party had started on Tei's side of the river, but our young friend was reasonably sober when he arrived. Liv had no objection to my letting the billygoat, Maita, loose to escape into the mountains where it belonged. But she came down the ladder with her snuffling piglet under her arm, insisting that she take Mai-mai along.

'You're crazy,' I whispered. 'Before we get away from this island, Mai-mai will be big and fat and cause havoc among other passengers on the ship.'

But Liv was not to be persuaded. And as we took off for the dark forest, she gave the stubborn piglet to our guide, as it bluntly refused to follow her on the leash. The guide, on the other hand, indignantly refused to carry the pig, so I carried it myself, together with the camera in a rug.

We bade farewell to nobody. They were all drunk, and, with the centipedes fresh in our minds, we knew we should be in grave trouble if Napoleon and his army realized we were about to escape.

We had not gone very far before we decided to wait for daylight, as there was no trail through the boulder-strewn hibiscus forest in the inner part of the valley. But at least we had escaped, and when the drunkards came across the river they were destined to find our pole cabin empty.

As soon as day broke, we continued inland along the valley bottom. Mai-mai screamed and shrieked and wriggled like a freshly caught salmon. I carried her in my arms, on my shoulder, against my chest, and inside my khaki shirt, but the piglet screamed no matter what I did. And now the sun began to bake us. She had to be taken out of the rug, although it made me hot carrying her in my arms.

When we finally crossed the stream, I dipped Mai-mai into the water with the intention of cooling her off a bit, but then she shrieked worse than ever. It was just not bearable. Our guide, who had walked in front, increased his pace and disappeared completely – along with the pack. And when we reached the foot of the Tauaouoho Mountains and entered *teita* grass almost as tall as ourselves, he was gone. Lost, as if he had sunk into the earth. We had to begin searching for the overgrown entrance to the mountain trail all on our own.

The Ouia Valley ended here, and we were to begin our ascent serpentine-fashion up the burning rock wall. There was not so much as a tiny tree to give shade, and the sun was baking hot on this low part of the mountainside, without a breath of wind. The dry *teita* grass let the stinging rays pass through everywhere. Mai-mai's piercing shrieks seemed to augment the burning heat, and I suggested meekly that we set the piglet free. But Liv protested vigorously.

'The poor creature will just fry to crisp bacon on the hot rock,' she explained.

And we advanced slowly, higher and higher, walking and crawling. The steep trail we were following was indistinct and grew even worse, until we followed nothing but some barely visible tracks in the sand between the stems of the tall *teita* grass. At last we came to some overhanging cliffs and sheer drops, and here we were completely stuck. I knelt down and examined carefully the tracks we had followed. They were those of a wild boar. We were lost.

To advance in the *teita* had not been so easy. But returning along our own tracks was worse. The long, bayonetlike blades were bent, cutting our skin like the edges of opened cans. By now, we began to fear that we had been led into a trap. The fellow who had disappeared with all our worldly property might be back at our heels with a posse of drunken natives. We had to hurry up. Hurry back to where we must have passed the right entrance to the mountain trail. Hurry to the plateau, where we could escape to anywhere. We were really cornered.

When we finally discovered the track, we were just about exhausted. The air stood around us like the mountain walls, hot and motionless. It was the inside of a baker's oven. The sun scorched our countless cuts and scratches, and all our pores were sealed with dust, sand, and perspiration.

The piglet was hot and damp in its short, thick fur, and gave us not a moment's peace. Its piercing screams cut through to our bones and marrow in this heat, and at times I felt a desperate desire to throw the wriggling thing over the cliff.

The wall rose high above us; we were not yet halfway up. We should never get all the way up without a rest, and to rest was impossible in the sun. The sand and the rock itself were burning hot, only the wide hats plaited by Liv and Momo from pandanus leaves saved us from sunstroke in the equatorial heat. It felt as if we were breathing the same sultry air over and over. We had to climb even higher, for up there somewhere we would surely find the breeze. Our fixed idea was to reach a strange rock formation high up in the cliff face. We remembered the place from our descent into Ouia. Somewhere up above us were two pinnacles shaped like twin trolls in a Norwegian fairy tale. Two big petrified sentinels overlooking the valley and the sea, with just enough space for a short tunnel between their straddled legs. And this was the only place in the entire mountain wall with any shade at all. We had been told during our descent that even when the air hung dead calm and quivered with heat everywhere else, a gust of air moved up through the tunnel straddled by the two stone giants.

An eternity passed from the moment we sighted these twin trolls above us before they seemed to grow and move nearer. The last stretch became too much for Liv. She stumbled continually and had to be fanned with the big hat. During this brief cooling procedure, even Mai-mai grunted contentedly. But when the fanning stopped, the little devil shrieked twice as hard.

It was an indescribable treat, a superb physical pleasure, when the trolls stood above us and we could dive into the dark, airy tunnel. The main rock face continued above us, but up here we could see that nobody was on our heels. We could stay until night fell upon the island, even though our guide had escaped with the provisions: the coconut milk as well as the baked tarn roots Liv had prepared for the crossing.

When our eyes grew used to the shade, the tropical valley below us seemed blinding white in the intense light of the midday sun. Mai-mai grunted peacefully and fell asleep in Liv's arms. The shade and breeze cooled and dried our skins, and, a few hours later, we felt an itching desire to continue our escape into the highland. We crawled back into the afternoon sun and resumed our ascent.

In a difficult spot, to free my arms for crawling, I had to put the wriggling, screaming pig with the camera into a sack formed by the blanket on my back. After a while, the noisy animal calmed completely down.

'It thrives in the dark sack,' I said, happy at having found a solution. But Liv insisted on peeping down into the blanket, and there the piglet lay as motionless as on a Christmas dinner table. She quickly fished it out, and, restored to a pig's full senses, it began to scream as before.

In this way, we reached the edge of the precipice. The inland mountain plains lay before us. The strong breeze had caught up with us during the last part of the climb, and now we had only one thing in mind: to head for the nearest mountain spring.

We drank from every spring and rivulet we passed. The first pool tasted better than iced champagne. Mai-mai shared our pleasures. We were hungry, but we did not want to descend into Omoa. We did not want anybody to know our whereabouts.

As night fell, we found ourselves in a gorge between the hills. It felt terribly cold. We shivered, and longed for a fire. While Liv used the last daylight to collect ferns for our bed, I struggled with two dry rubbing-sticks until I was exhausted. The wood grew black and smelled like sweet incense, but that was all. With numb arms I gave up, but cold and despair made me start again. Smoke. Immediately Liv was at my side ready with tinder, but I could hardly hold out any longer. We should just have to freeze under the open mountain sky. I was longing for a match. Just one match. Or at least Tei's flint. This wood was not right. Then a spark was born in the tinder, a little star. Liv blew with the utmost care. As living fire came out of my sticks and flamed up from the tinder, I felt as proud as the owner of Aladdin's lamp, able to light up the invisible surroundings and create a pleasant temperature in the chilly mountain air. To give us a feeling of security against centipedes and wild dogs, I lit a ring of small fires around us. Liv had made a soft green berth in the middle of the trail, the only place where we could find smooth, open ground.

Then we lay down in the gorge, covered by our blanket, enjoying the free view of a million tropical stars and constellations rotating slowly above us. A global compass framed by rugged mountain silhouettes.

But Mai-mai screamed so angrily in Liv's arms that neither of us could sleep. We had discovered that the piglet was not a he at all but a she, and I took the opportunity of renaming her Siren, which seemed the most fitting name.

After a while, even Liv recognized that Siren's company was unbearable. I happily suggested tying her somewhere out of audible range, and Liv consented, on condition that the innocent little devil should get our only blanket. I walked far down the trail and tied the blanket, a sack with Siren inside, to a stone. From now on, we heard only the night wind in the trees, and we fell asleep with flickering fire on all sides.

In the middle of the night, we awoke to hear and feel the drumming of heavy hoofs. I was stiff from cold; all the fires had long since burned down. But what woke me up in less than a second was the sight of two wild horses coming straight toward us, silhouetted in the moonlight. With long, flowing manes and tails, they came chasing through the gorge at a mad pace.

I sat up with a terrible yell to scare the horses, but too late. The one in front did not manage to stop and, panicky from fear, it made a terrific leap over both of us. The second horse managed to skid to a halt then reared and galloped off in the opposite direction.

Both of us leaped to our feet and poked life into the fires, whereupon we threw on more branches to see and be seen, and to get our blood back into circulation.

'Siren,' was Liv's laconic comment.

No doubt whatsoever. The little piglet had been disturbed in its sleep by the sound of big animals coming along the trail. Starting to scream and dance about inside the rug, like a ghost in the moonlight, she had scared the two wild horses out of their wits. The little scoundrel had unwittingly given us a narrow escape.

Empty stomachs made us surrender next day. We descended into the Omoa Valley to look for food. The mountain trail ended by the water's edge, and down there, to our surprise, we found Willy sitting on the beach as if waiting for our arrival. With his usual calm smile, he told us that our other blanket, with all the stuff our guide had carried across the mountains, was safe in Willy's own house. He had seen the man coming down the trail and had taken care of everything he carried.

We had almost forgotten that we had a friend in Willy. He was different. He had never sided with those who caused us trouble.

Grateful to Willy, and restored by a sturdy corned-beef meal in his bungalow on the beach, we left our gentle host with all our recovered property. We had only vague ideas about what to do. First we headed for our old friends Pakeekee and Tioti.

On the road, we were stopped by a man in a straw hat and a loincloth, who offered me a bargain. If I let him have Liv, he would give me his wife and four children in exchange. He opened his arms as if he were measuring a barrel to make me understand that in size I was getting the better part of the deal. He seemed surprised when Liv and I jointly turned the offer down.

We hurried on and found Tioti. He was pleased to see us and even pleased to be presented with Siren, who was hoarse from screaming but calmed down and started grunting like other pigs when she was left to run loose among Tioti's chickens. Tioti had a solution, as always. When the village was sleeping we should proceed to Tahaoa, a not easily accessible beach with white coral sand at the foot of vertical cliffs near the north cape. Tioti had taken us there to collect shells once before. If we remained there, he would come and bring us food from time to time. And when the schooner arrived, he would hurry over and let us know.

XV

Cave Dwellers

WHEN THE VILLAGE SLEPT and darkness reigned in the forested valley, we sneaked down to the open, starlit coast and began to fumble our way with our few possessions over and along the boulder-strewn foot of the seaward cliff. We headed for Tahaoa. We knew the way.

We had to climb over enormous fallen blocks and jump from one to another with the surf frothing against them and between them and the vertical cliffs above our heads. Sometimes when a big surf came breaking against us and we were afraid of being swept into the cauldron, we had to climb up on the biggest rocks to wait for a safe moment in which to continue along the rock fall. Gradually the rock fall ended, and we went along a narrow shelf of black lava. The action of both water and former volcanic activity had created the most curious formations, with caves and natural bridges. Sometimes we heard a great sucking noise and a lion's roar under our feet, and a geyser of water would shoot up into the air.

About noon, we rounded a corner of the precipice to which we'd been clinging, and jumped down on a sublime beach. I could hardly believe my eyes when we'd come here before with Tioti and his wife. Blinding white sand and scattered blocks of coral, white as snow, ran for almost a mile along the foot of the cliffs and a stone's throw out into the sea. The only coral reef on Fatu-Hiva. The only one on any of these cliff-girded volcanic islands of the Marquesas. It ran as a slightly submerged beach shelf and created a wide belt of calm, clear, green water, where water-worn rocks and reef emerged everywhere, enclosing pools and shallow channels. Behind this bright belt the gray overhanging rock rose sky high and left room for nothing else but a very narrow strip of grass with a few tall palms and clusters of shrubs.

The place was at the same time breathtakingly beautiful and yet forbidding and menacing. The overhanging rock walls closed the entry into the island. Visitors like us, coming in from the side curtain, remained tightly squeezed between the sky-piercing wall and the endless sea.

Here too was an imminent threat of stone fall. The only secure place was a shallow cave in the rocks. This was where we headed with our modest belongings.

Tahaoa was a sunlit world of its own, completely unlike the black beaches of boulders and lava sand typical of this volcanic group. The light was intense. On arrival we were at first blinded by the reflections from hundreds of glass-clear pools in the reef and the glittering snow-white coral sand. Sand ground from billions of shells and white bits of coral. The surf from the open ocean sent white cascades high into the air along the outer edge of the reef. Only friendly ripples managed to move in over the rim and across the reef itself, to form small rivulets of fresh ocean water running in and out of the labyrinth of ponds and channels.

With patches of vivid lichenlike coating in red, yellow, and green, the shallow bottom of white coral and black lava showed up everywhere to form an outdoor aquarium like none we had ever seen.

Our friends the hermit crabs were busy on the sand and in the ponds. Shells everywhere. Empty shells and shells occupied by these crustaceans or by the drowsy original owners, who exhibited a wide variety of architectural ideas in their house design: conches of all sizes, leopard shells, turban shells, bubble shells, tooth shells, sea ears, and all sorts of common sea snails, clams, and other mollusks in the profusion only tropical waters can offer.

Sea urchins, starfish, and crabs posed motionless or crawled about among colorful sea anemones in a garden of algae. Painted more colorfully than the rainbow and molded into every imaginable shape, members of the fish family wriggled and paddled about or hid behind marine foliage, some thick as pears, some thin as a pin, some twisting like worms, some flat like pancakes. Each kind differed from the others in appearance and behavior. Each species seemed to be granted the patent for a special useful organ that the others were denied; ingenious devices for propulsion, for propagation, for escape, or for attack; for the survival of its own species.

Many of these artistically decorated creatures were so richly embellished, seemingly for no useful purpose, that it was tempting to look around for the missing spectators for whom all this beauty was designed. And if the purpose was not beauty but some function as yet unknown to us, then it was all the more reason to look around us for the talent behind the purposeful design.

Some evolutionary force had with extreme luck or supreme intelligence composed molecules and genes and then rearranged them to change single-celled plankton into all these mobile creatures with eyes, tongues, claws, tails, and fins. The success of this enormous creativity was visible everywhere on the reef and more real than the rock wall that towered above us. It was there and yet as invisible as instinct, as gravity, as magnetism, as a hundred thousand volts.

This was where Tioti, the sexton, had found the large conch he blew in his futile attempt at luring some of the Catholics of Omoa into Pakeekee's little Protestant bamboo church. Had he blown his conch here, he could have gathered people of all religious creeds to a temple with sky for a ceiling and decorations by that great designer we celebrate indoors in temples built to our own taste. There was a feeling of Sunday every day in Tahaoa. Man had let Nature rest as it was left by the Forces of Evolution on the Seventh Day.

Tioti had promised to bring us fruit at intervals. There was nothing but seafood in Tahaoa. On arrival we put our bundles into the cave and waded into the shallow reef to get acquainted with our new surroundings. I put my nose into the water to see better. Small red fish instantly registered danger and hurried away, fast as arrows, to seek camouflage over a red plot of marine growth, where I could hardly see them. Did they realize how smart they were? Who had taught them this trick? I tried to grab some tiny blue fish. They hurried in between the bristling quills of the poisonous black sea urchin. Somehow these fish must have realized that this was a secure fortification against larger enemies, so they shared the protection of the needle-sharp bayonets of the sea urchin, whose poisonous, fragile barbs would break off inside any wound it made.

Even the pearly shells of seemingly brainless clams closed their gaping doors so securely that no fingers could force them open. Inadvertently I almost stepped on a small octopus, that quickly shot backward, sending out a black smoke cloud hiding it in its retreat. These 'stupid' lumps of pink flesh were shaped like rubber bags and fumbled along the bottom with their long tentacles in a clumsy way. But if frightened, they stretched out their eight limbs like a teacher of gymnastics and shot backward by pressing water out of their built-in jet propulsion system, so fast that they were gone before anyone could guess at their whereabouts behind the cloud of pitch-black ink. Their relatives, the ten-tentacled squids of the open sea, can shoot backward through the water so fast that, like flying fish, they can unfold lateral flaps like magic cloaks and sail through the air out of reach of marine pursuers.

It was easier to grab a hard-shelled crab who felt so secure inside its armor that it merely remained straddling a stone, waving its antennae with claws lifted like fossil boxing gloves, and eyes on mobile shafts scanning in all directions like a mechanical toy. Once in my hand, it tried to pinch me until I put it back on the stone, and then moved away like a robot.

All these peculiar creatures mated in their own way and with their own breed, to multiply or, rather, to maintain their own number. If the number increased drastically, the machinery would stop were it not for the interference of the other species. Liv, as a former student of social economy, was amazed at the fact that a common fish might spawn roe equivalent to 100,000 babies, of which only two would survive to spawn the next generation. How did the sharks, the pelicans, and other predators know when to stop catching more, once the allotted ration of 99,998 was consumed?

I had no answer, but we could both see the complex balance between the various inventions for both assault and defense. Shields and solid armor with jointed and hinged limbs, along with spikes and thorns for self-defense, were common and ingenious devices for fast escape and camouflage. Daggers and spears, saws and hooks, tongues and pincers, snares and traps, suction disks, electric-shock devices, and damaging chemicals formed part of the innumerable facilities this marine society had obtained as genetic birthday gifts. All to guarantee mutual survival and balance in the eternal, precise clockwork Nature had created.

Tahaoa was a world of its own, especially at night. Lonely and empty, with the bare wall behind us and the ocean void of surface life meeting the star-filled vault of heaven. Even the birds slept. As night fell the first day, we stood as if petrified, looking at the starlit sea, introducing ourselves to these new surroundings Then we crawled into our rock shelter. We were the new lodgers. There was no sign of others having ever slept here before. We had brought nothing but the tiny tent we had tried to use the first night under the coconut palm in Omoa, our two blankets, the big knife, and two palmleaf baskets filled with fruit and baked roots. Outdoors, the tent was in as much danger from falling stones as from falling coconuts, but the rock shelter had a ceiling that would resist anything. No insects would devour these walls, no boar would make them sway.

The floor of the cave was covered with a jumble of large and small boulders as smooth as eggs. I rolled the biggest boulders out front to form a sort of barricade against the sea, and removed even the pebbles to make a smooth place for a bed. There was white sand beneath them. As I rolled a large boulder over to reinforce the rampart from the outside, a giant, glittering moray eel twisted angrily about in one spot like a fat, green-and-black-speckled serpent. Finally making up its mind, it slid away between my legs into the pools of the reef.

I had not known before that these nasty marine beasts would move up onto the land. Surely at high tide the mouth of the cave must be awash. The boulders outside our rampart were glittering wet underneath, and seawater seeped into the deeper depressions. Nevertheless, the moray had wriggled seaward over perfectly dry stones, like a short and stubby boa constrictor.

We had learned to fear these snakelike beasts more than any shark. When we'd gone fishing with the natives, we had seen how the men pulled sharks right onto the rim of the canoe and then just hit them on the head with a heavy club. But if the hook was swallowed by an evil-eyed moray eel, and it came up with its awl-like teeth gaping, then they would yell in excitement and never pull it in until they had pierced its head repeatedly with their three-pronged spears. The thin, sharp teeth of the moray could cause a poisonous bite, and according to our island friends, big specimens could tear off a man's arm. They were often thicker than a human leg, and once in Ouia I had stared into the snake-like eyes of a colossus outside an underwater grotto, the weaving forepart of its body as thick as my thigh. Several islanders claimed to have seen a moray with a body as stout as the trunk of a coconut palm. They were undoubtedly referring to truly giant specimens, which might have passed for veritable sea serpents.

I was more careful about where I put my fingers and feet when I resumed rolling the boulders over. There could be other morays under any of these stones. In the dim light from the night sky, we completed fitting out our new residence. Liv had gathered as much grass as she could find, and with it she covered the floor of the cave. We crawled to bed, rolled up in our two rugs, and slept until the intense reflections of sunlight on the reef woke us up in the late morning.

Drowsily I sat up and looked around. Rocks and water. Liv lay wide awake, her head resting on her hand, and gazed out over the sea, her expression not revealing any of her thoughts. She had by now taken a lot of beating, and had never complained. She had never said that this was all my crazy idea, or asked why all this had happened to her. Never said that she wanted to go home. Her home had been Fatu-Hiva. Wherever we lighted our campfire, she had adjusted herself to the conditions.

As for myself, I did not quite know what to think anymore. We had been defeated – and yet not quite. We were refugees, but were still as free as the frigate birds sailing about over the reef. We had come to live close to nature, and we admired nature more now than ever. Yet things had not worked out the way we had expected.

We had tried to live deep in the jungle, in the open hills, and under the palms by the sea. We had succeeded in each place for a certain length of time, but something had always turned up to put a spoke in the wheel. Now we were to start again as cave dwellers. Cave dwellers on a beach. We were squeezed in between the black toes of a hanging rock wall and the licking tongue of the sea. Fresh water came dribbling down the rock face, and there was enough food to pick from the trees and bushes and to catch in salty pools to keep us alive. Yet this was not exactly what we had dreamed about when we packed our suitcases for the long journey back to nature.

I crept out and warmed up in the sun. What an enormous ocean! And there was only a narrow strip of land at our disposal. I looked up. I hoped that goats and booby birds would mind their steps up there, otherwise eroded rock would tumble down upon us.

This was not a place to remain for life. Not a place to raise a family in. Liv could be blessed with a baby at any time. Nature alone decided. Her motherly instinct had been revealed in the way she had fondled Siren.

Liv and I did not have much to say as we set to work to organize our new life. There was not much to say about a future on the Tahaoa beach. We had little to organize, but we had to construct a sheltered fireplace and pull in under the overhanging cliff as much driftwood as we needed to keep our hearth going. No table, no bed, no bench were required here. We still had our coconut cups, our bamboo spoons and beakers. We needed no door to keep boars or mosquitoes out. If rain came from the direction of the sea, we could hang our tent down over the cave opening.

I tried to climb to the top of the shortest coconut palm. It was too high. We'd be satisfied with the few nuts that had fallen to the ground until Tioto came. A mediocre climber could easily crack his skull, and there was no *taoa* available to mend it.

Liv was wading, her pareu held up to make a pocket that she was filling with edible mollusks. I joined her. We agreed that the reef was the most wonderful aquarium we had ever seen. Otherwise, we did not say much. I warned her to be on guard against octopuses and moray eels, and not to step on sea urchins. She said she would. That was all.

As we sat on the boulders washing our food down with coconut milk, I proposed that perhaps we could find some seabirds' eggs. She agreed. Perhaps we could. We finished our meal in silence.

I tried to read Liv's mind, and came to the conclusion that, if she was thinking at all, both of us had to be thinking along the same lines. We had started this experiment with the same ideals, the same dreams. We had lived through exactly the same experiences, seen the same wonders, suffered the same disappointments. We were not as green as we had been when we came. Both of us had hardened a bit. Both of us had come to realize that we had been too egocentric and had almost ignored that there were other people in this world. We were not so stubbornly sure of all our own visions and calculations. Unpredictable obstacles had thrown us off what we had thought to be an open road. Now, in fact, we were not on any road. We had to digest all our unexpected experiences before we could set ourselves on firm ground again.

There was ample time for thought in our cave. For rethinking. For examining our current feelings toward the civilization from which we had sought to escape. To crystallize our notions as to what had come out of our experiment of turning back to nature. This radiant beach was clearly a blind alley.

For several days, we felt uncertain of ourselves. We spent much of our time bathing in the clear pools or wading on the reef, catching fish, crabs, and other crustaceans with our bare hands. At low tide, it was difficult for them to escape from the landlocked channels and pools. And there were delicious sea snails and other edible mollusks that did not even attempt to escape. We picked them as a farmer picks tomatoes.

Only once did Tioti and his wife visit us. When they came, however, they brought with them a huge load of fruit, nuts, tubers, even poultry. We hoarded everything in the cool innermost nook of our cave. Our visitors left, and one day followed the next. We rose with the sun and crawled into our cave when it set. And made sure that the embers of our fire never died out.

Most of the time, we just sat in the shade of the rock shelter, scanning the horizon. Once more we sat as if shipwrecked on a reef, with an ever-growing desire: to see black smoke or white sails on the faint division line between the blues of sea and sky. Among the thousands of whitecaps that came and went, we hoped to detect one that grew into approaching sails.

'What do we do if we see a schooner?' Liv asked one day, after we had been sitting staring at the horizon all morning.

'We hurry to Omoa, of course,' I answered, surprised. 'And if it arrives at night, Tioti will come and fetch us.'

'Why?' she asked, and gave the answer herself. We were killing time now. We were getting to be like the people in Omoa who just sat outside the walls of their homes, waiting for the coconuts to fall.

It felt like a relief when we both realized that we had the same idea: this experiment had come to an end.

Neither of us would for a moment have missed our experiences on Fatu-Hiva. Liv stressed that she would not have missed our time on this island for all the treasures in the world. That went for me, too. I would not have skipped a day of what we had behind us. When the schooner came and we could get away, we would carry with us lessons that would last a lifetime.

When, for the first time, we conversed openly about an ultimate departure, it felt as if a coat of ice had melted at the bottom of our hearts. The play of sunlight on sand and reef was suddenly as pleasant again as on the day we first arrived. We were not prisoners. We were not wedded to Tahaoa. The world we once knew was distant, but still there, even though it had for so long been far from our thoughts. Our parents also.

For the first time since we had come, we became lost in reveries about how it would be to see our parents once more. When we had boarded the train that cold Christmas Day and took a seemingly joyous farewell of family and friends, inside ourselves we were bidding them a sad good-bye. We had not been at all sure that we would ever come back to civilization again. In fact, when the French island administration had demanded prepaid return tickets before they would let us disembark in Tahiti, I had been obliged to comply with the regulations. At that time a prepaid return ticket seemed a waste of money. We could use those tickets now.

'Do you realize,' I said to Liv, 'if things had worked out differently, we should still have left Fatu-Hiva. If we had found that man could rid himself of all modern problems by going straight back to nature, we should have been pestered by guilty consciences until we went home and told others – '

Something of the insect is within us. An ant has invisible ties to the anthill. A bee finds no satisfaction in hiding from the hive and enjoying its harvest in solitude.

'But do you know something?' Liv interrupted. 'If things had worked out the way we had expected, we still couldn't go home and recommend a mass migration back to nature. Think of our map!'

She reminded me of the map on which I had crossed out continents and islands one by one before we drew a ring around Fatu-Hiva, the only dot on our planet where there would be a chance to live in the wilderness of our time. Nowhere else were there abandoned gardens where food could be harvested as in the days of the mythical parents of mankind. I was still sure there was no alternative spot on earth. Tahiti would have been impossible. On all the other islands every piece of fertile land belonged to somebody. The kind of food modern man is able to digest no longer hangs on trees in no-man's-land.

The world itself had changed as much as man himself since the early days when he started his fight for independence from nature. There was no road back to our lost beginnings.

'There is nothing for modern man to return to,' I admitted reluctantly. I said it with sadness, for our wonderful time in the wilderness had given us a real taste of what mankind had once abandoned. And what mankind was still struggling to get ever farther away from. And in his attempt to build a better world on the Seventh Day, man was still in the middle of a long road. Far from his point of departure and still with no sign of a sensible destination ahead. We were looking for progress, but should not fool ourselves into thinking that just any road ahead is that progress.

We had come to Fatu-Hiva full of contempt for twentieth-century civilization, convinced that man had to start all over again from scratch. We had come to take a critical look at the modern world from the outside. Now we were sitting in our cave with a rock wall over us, staring into the empty blue, waiting for the means to turn back, not to nature, but to civilization. We were milder in our judgment. We had seen that without Willy's mosquito netting we would have been driven out of our wits in Fatu-Hiva's jungle and ended up with elephant legs. Without Terai's ointment on Hivaoa, we would have ended up with no legs at all.

Yet, we had not gained full confidence in modern civilization. We had seen how simple life could be, how perfectly relaxed and intensely happy a person could be deprived of the countless items we struggle to get access to when we want to live like others in a city block.

We felt an urge, an inconvenient need, to return to civilization. But we did not want to be a single step farther from nature than life in our part of the world made necessary. Life in the wilderness had filled us with well-being, given us more than the city life as we knew it had ever been able to give us. Never had we seen people at home, not even our relatives, laughing so readily and reflecting such a free and healthy spirit as Tahia-Momo and Tei Tetua, who were poorer when it came to property than any other people we had ever met. Deprived of all the benefits of untold centuries of cultural progress.

I had no sooner reminded Liv of this than she corrected me. We had met the little Frenchman in the funny shack on Hivaoa. His shack was full of inventions and devices to make his life easier.

He lived in a world of books. He had created his own kind of civilization. He, too, was indeed a happy man. And he was neither a child of nature nor an illiterate. His recipe for happiness was to be found between the walls of his tiny home as soon as he came in, and outside whenever he went out. He carried his own paradise around with him, inside himself. As its ever-present source. If the environment around him had helped him to find it, the secret could be crystallized in one word: simplicity. Simplicity had given him what millions of others searched for, not complexity and progress. The old man's requirement was a shack in a vegetable plot. Neither a cave in the wilderness nor a castle in a city.

Once we had admitted to ourselves and to each other that we wanted to leave our cave and return to our own world, we sat until dusk, eating hot clams and discussing civilization the way we saw it from a distance, with all its blessings and all its adversities. When the western sky turned as red as a royal carpet in the wake of the retiring sun, a violet veil was slowly drawn up across the other half of the sky. Liv scraped away the thick cushion of ashes from our precious embers and blew new life into the dormant fire. For the first time since we came to the beach, we sat on the boulders long into the night talking until the same red and the same violet changed positions in the sky and the stars began to fade.

There was no schooner the next day, nor the next. We strolled on the beach among dead shells and rattling hermit crabs, talking. Always on the lookout seaward. A distant wave breaking white against the blue sky, a killer whale leaping as from a springboard, a white bird on the horizon, nothing more was needed to put us on the alert. A schooner must not pass by unnoticed. We would climb up on big rocks to see further. We had a view of half the ocean within our horizons.

The ocean seemed to overflow and enter into my very soul during those days in Tahaoa. The salty air filled every breath, and at high tide the smallest ripples across the reef ran right up to the mouth of our cave. Hermit crabs crawled across our barricade and stole our food like rats. Fish were begging like dogs at the foot of our table. All we chewed and swallowed tasted of seaweed or the salty ocean.

Sitting staring out into the blue Pacific, which ran without a clear transition into blue space, I felt that the ocean was immense, endless, bottomless. There was something beyond human comprehension about its immeasurable size, since the Amazon, the Nile, the Danube, the Mississippi, the Ganges, all the rivers, floods, and sewers in the world could enter into it ceaselessly without the surface level ever changing an inch. All the running water in the world heads for the ocean, yet it only churns its currents around as in a witch's cauldron, its surface calmly rising and sinking with the tides, ignoring the fact that it ought to overflow, that it ought to run slowly into our cave and start rising up the cliff wall behind us. All the rain and all the rivers have no effect on its level. All the mud, the silt, the rotting vegetation, the carcasses and excrement of animals from sea, air, and land that have washed into it since the days of the dinosaurs and the first life on earth, have failed to pollute it and have left it perfectly clean. For man, the sea and the sky have been the two symbols of endless dimensions and permanence.

I knew very well that the ocean would never overflow, because all the rivers that run into it contain precisely the same water that has evaporated and come down again as rain. Nevertheless I was sitting there in front of my cave – I who had the mentality of a caveman, of a medieval European, of most people in the modern world – sharing the feeling that the land was man's domain but that the ocean was part of space. Little did I suspect that I should in later life get a completely different impression. That I should get to fully understand that the ocean was the pulsating heart of our own living planet. That it was possible to cross it by raft. On bundles of reeds.

All this I was to learn much later. Now I was sitting watching the ocean as something with no beginning and no end. Something man could never threaten. Suddenly I felt as if the blood rushed to my head and I hurried up on a rock to focus my eyes on one point on the horizon.

'Liv,' I shouted. 'A sail!'

'Where?' In a second she was at my side. 'Yes, I can see it!'

With beating hearts we gazed for a moment at the majestic white sail of a schooner at first fixed to one spot, then slowly growing bigger. It was coming from the direction of Tahiti and heading for Fatu-Hiva.

We jumped down onto the sand and ran for the cave to get the camera and the machete. We lost no time over anything else, and then ran as fast as we could along the white beach, up the black lava boulders, leaping and climbing from stone to stone along the foot of the cliff toward Omoa.

The schooner was just rattling its anchor chain to the bottom as we jumped down onto the grass of the sunny bay. Everybody was there: Tioti, Willy, Pakeekee, Ioane. Everybody was smiling sadly because we were about to leave. Now we liked them all. They helped us carry all our heavy boxes of stones and bottles down from our jungle hideout and drag them into the lifeboat.

We hated leaving. We hated going back to civilization. But it was something we could not resist. We had to do it. We were sure then, and I still am, that the only place where it is possible to find nature as it always was is within man himself.

The first thing I searched for when I opened my moldy suitcase was our return ticket from Tahiti to Europe. I found it.

'Liv,' I said, 'one cannot buy a ticket to paradise.'

These were the last words in the book I wrote on my return from Fatu-Hiva.

XVI

The Seventh Day of Eternity

'ONCE IN A WHILE you find yourself in an odd situation. You get into it by degrees and in the most natural way, but, when you are right in the middle of it, you are suddenly astonished and ask yourself how in the world it all came about.

'If, for example, you put to sea on a wooden raft with a parrot and five companions, it is inevitable that sooner or later you will wake up one morning out at sea, perhaps a little better rested than ordinarily, and begin to think about it.'

These were the first words in the next book I wrote. Ten years later.

I then referred to a morning I woke up in a bamboo cabin that was rolling about in an eternity of ocean on top of nine lashed-together balsa logs. The golden cabin walls around me were of plaited bamboo, like those of our first jungle home on Fatu-Hiva.

But this time the wall next to my side was wide open, and I had an unimpeded view of a vast blue sea with hissing waves, rolling by close at hand in a relentless pursuit of an ever-retreating horizon.

Somewhere far beyond that retreating horizon, somewhere far ahead of our clumsy bow but so far away that our morning sun had not yet risen there, lay Fatu-Hiva. Some of the whitecaps that danced along with us and pushed us forward would end up hitting the rocky walls of that island, and tumble in on the boulder beach of Ouia Bay. Perhaps our pole cabin was still standing. In a few hours the morning sun would rise there too, and old Tei would tumble out and look in our direction without the slightest suspicion that I was there, arriving slowly in the wake of the sun. Repeating the voyage he had told us about in his songs and legends of the seafaring god-king Tiki.

We had named our raft after that immigrant pan-Polynesian deity. We had added *Kon* to the name, since the vessel sailed from Peru; *Kon* had been the pre-Inca name on the Peruvian coast and *Tiki* in the highlands for the pan-Peruvian god-king from Tiahuanaco, who had embarked with all his followers on his final departure into the Pacific.

It was on the *Kon-Tiki* I woke up one morning a bit bewildered to find myself on a raft dancing among ocean waves. I had sworn never again, never ever again, to go on a deep-sea venture in an open boat after my experiences in a dugout canoe and the plank-built lifeboat in Polynesia.

And yet here I found myself among those waves again, this time on a mere raft on her way out into the world's largest ocean, thousands of miles from land and outside all shipping lanes.

I had reason to ask how I, known to myself and all my friends as a notorious landlubber, had felt the urge to embark on a balsa-wood raft on the open coast of Peru and sail on a course for Polynesia, 8,000 kilometers away. To get there we had to drift and sail a distance that, as the crow flies, equaled that from New York to Moscow. Of course, everybody had reason to think I was crazy.

None of us on board had seen balsa wood before. None of us had the slightest experience in sailing. The experts said the balsa logs would sink after two weeks. I had learned to swim when I fell into the river in Tahiti. But there was nowhere to swim to if the logs sank. My five friends on board knew I was far from a dare-devil. They assumed I would not take an unreasonable risk. And that was probably why they trusted my calculations. To me, my deductions seemed simple:

A: Certain cultivated plants had been brought from South America to Polynesia by pre-European people.
B: In South America there were no boats in pre-European time, only balsa rafts.
A + B: Accordingly, a balsa raft must be able to float from South America to Polynesia.

Eventually time would show that this reasoning was correct. Some experts said that the *Kon-Tiki* voyage proved nothing but that Norwegians were good sailors. They missed the point. They stuck to their theory that the pre-Incas on the coast – like the Incas of the highlands – were landlubbers. If so, the *Kon-Tiki* voyage proved that even landlubbers could have drifted from Peru to Polynesia on balsa rafts, and planted sweet potatoes upon arrival.

To me the voyage meant something more. I got to learn both the amazing quality of the favorite transport vessel of the Inca, and the friendly attitude of the ocean to this type of wash-through watercraft. The 101 days we spent on board the *Kon-Tiki*, while wind and current pushed and pulled us across the 8,000 kilometers from the mainland to the islands, gave us abundant opportunity to understand that a raft of balsa logs, lighter than cork, is far superior to any other open watercraft in carrying capacity and security for crew and cargo. The many great and changing civilizations all along the 2,000 mile long desert coast of

Peru had stuck to this specific type of craft for millennia for good reasons. Not, as we think, because they knew of no better type of vessel, but because they knew there was no better type of vessel for their special needs. On rivers they used both rafts and dugout canoes. In the sheltered waters of the archipelago along the coast of Chile they built sinkable vessels with ribs and hull. But to ride the towering surf and venture among the reefs and rocks of their own totally open coastline, nothing was safer and better suited than shallow-drafted, bump-proof, wash-through rafts of buoyant balsa logs or compact bundles of *totora* reeds.

We are amazed by their mountain-size pyramids, and by all their other feats of monumental architecture. By their incredible aqueducts and complex hydraulic engineering. By their plant and animal domestication, which involves highly advanced agricultural systems and navigable canals that run for hundreds of miles through the coastal deserts. Their unsurpassed techniques and skill in tapestry weaving and ceramics. We are no less impressed by their knowledge of surgery, metallurgy, astronomy, and social and military organization. Symbolic and realistic representations of their daily life may be found in their elaborate sculptures, painted on pottery, and woven into cotton fabrics. Among all pre-Inca cultures along the coast, the dominant motifs are deep-sea fish and fishermen on large and small reed boats. Royalty is shown traveling on the upper deck of roomy oceangoing vessels with a double stern and rows of people and water jars packed below deck.

Today, with their former kingdoms collapsed, conquered first by the Incas and then by the Spaniards, the poor fishermen along the coast paddle out to sea far beyond the horizon in tiny one-man reed boats. They no longer build majestic reed ships or monumental pyramids. But when we reconstruct the golden ages of pre-Inca Peru, we have to add navigable ocean vessels to the long list of advanced cultural achievements typical of the entire coastline prior to the descent from the highlands of the nonseagoing Incas.

If any of us on board the *Kon-Tiki* had been prepared for disaster by the unanimous warnings from sailors and scientists before we left, we all began to relax after the first week in the choppy sea of the Humboldt Current. The behavior of the raft, which always rode on top of the waves and never became water-logged, gave us a feeling of complete security and merry excitement.

Kon-Tiki introduced me to a different ocean than the one I thought I knew before. As the months passed, day and night face-to-face with the waves and the ocean as a dancing partner, I began to feel that the ocean was a friend. It was more. It *was* the mother of all life. It had been the life supporter and food provider for thousands of years for all the maritime cultures along the coast where we had embarked. The Incas had organized relay runners, who brought them fresh fish from the coast into the highlands. When the first Inca descended to the coast he worshipped the ocean and called it Mama-Occlo, Mother Sea, while he honored the sun as the primeval father. According to the Incan historians, Kon-Tiki, the founder of their royal lineage, had taught his ancestors that through him they descended from the sun. It must have made sense to the Incas that the god-king Kon-Tiki in the end sailed on course for the sunset. We too saw the red sun sinking into the sea straight in front of our bow every night.

At night, when the tropical stars twinkled from the black sky, and a myriad of phosphorescent plankton twinkled back from the sea, we too felt like gods on a flying carpet in the universe. There was nothing but us on the raft, surrounded by darkness and stars. We could well understand the pre-Incan artists who depicted Kon-Tiki and other bearded and hook-nosed deities as traveling on serpents undulating among the stars, symbolically pulled along by bird-headed men while they themselves enjoyed pulling in fishlines that had hooked rays, sharks, or sea monsters. We too pulled in giant fishes. On board our balsa raft, real life became like a fairy tale. After all, this real life was a fairy tale. How else could one describe the fact that the twinkling plankton in the sea around us were the ancestors of the six of us who sat above water and looked down upon them? Creatures like them were the progenitors of all life on earth.

The plankton we saw from our raft would never develop into human beings. Not even into whales and dinosaurs. Some of them were pelagic eggs of fish and crustaceans and would quickly develop into the same kind of creature that had spawned them. Their life cycles were predestined to end in the same mold that had formed their own species. Eggs of a gray herring would become gray herrings. Eggs of a black sea urchin would become black sea urchins. Spores of a green alga would become a green alga.

Not so with the first plankton to appear in the once-virgin ocean. There was more power in the plankton that was formed out of dead matter when life began. At the time when the most primitive forms of life began to drift about in the sunlit layers of the water, some were endowed with the ingenious mechanisms for transformation and evolution. Some, but not all. There was plankton then as now that forever continued to drift about with neither the urge or ability to become more than single-celled particles of immobile organic matter. Some, however, began to split and add more cells to their simple bodies. One group began to grow roots or other mechanisms that enabled them to grab hold and stop their helpless drifting. They turned into algae, ferns, cacti, and majestic coconut palms. As others grew, they began to wriggle and struggle until muscles developed and limbs and organs emerged that helped them choose their own roads independently of winds and currents. While descendants of their own kin continued forever to drift helplessly about in the sunlit surface layers, some began to dive or crawl up on the rocks to feed on the algae, the ferns, and the greenery. They grew fins and wings and legs, until their descendants included every mobile species from whales to bees, from mice to elephants.

How that happened is a true-to-life fairy tale that still causes controversy among biologists. It was magic to early man in the wilderness. It was the act of God to the scribes of the Bible. It is a puzzle, but of little concern to most of us during our daily life, when we are surrounded by city lights and the drone of cars and radios. But it was a most important part of our life and experience on board the *Kon-Tiki*. Sailing on a raft in a black night through an explosion of blinking stars and plankton, when the giant suns of a billion solar systems seem the size of the twinkling dust in the sea from which we descend, our horizons widen. The stars are far away indeed. Big becomes the world of man, and long our genealogy.

Telescopes and microscopes are not needed to observe that evolution is acted out quickly inside the womb of pregnant creatures. And a larvae crawling with a hundred legs can wrap itself up and crawl out of obscurity after a few weeks' hiding, to fly about as a six-legged butterfly with four large wings and the most sumptuous colors. That is natural. Because by definition everything nature does is natural. Again we fool ourselves with our own words. If anything is supernatural, that is what nature does. The only thing that is natural, that is what man can do to nature. Cut it to pieces. Kill it. Dissect it and find out how chromosomes and nerve systems work. Graft and transplant. One day we may even be able to imitate nature and create life the way nature did before it evolved us. But at the very best, we will never get beyond the stage of plagiarism.

The creation of the world of stars and atoms is a fairy tale if there ever was one. Logically, creation is totally impossible. But logic is the product of the human mind, which again is governed by a brain with a limited number of channels programmed for us by nature. If there is a beginning, something must have begun it. Lying on our backs on a rolling raft watching the stars circle right and left rhythmically around the mast top, we on the *Kon-Tiki* pondered the mysteries of life as people of all cultures have done since mankind developed a brain. We found no solution, but that we live in a fairy tale world and carry heaven and hell within us. There is no other space for them.

Darwin and the early scribes of the Bible tell the same story in different words. The sages who wrote Genesis told it in terms that made sense in their time, Darwin with a vocabulary that makes better sense to us. The concurrence of the two versions is curious and hardly a coincidence. How could the old Semitic emigrants from Ur in Mesopotamia know that life on earth began in an ocean that once covered all land? And that fish came before the fowls in the air and the beasts on land, and that we ourselves were the last of all species to appear?

Darwin had read Genesis, but the authors of Genesis had also seen what Darwin saw. Darwin made more use of his eyes in the wilderness than looking through his microscope at home. The authors of antiquity had also seen the wonders of nature that inspired Darwin. Nature inside and around man was there unchanged since philosophers began to think about the origins and growth of flora and fauna. The mysteries of fossils and fetus. Darwin had perhaps inherited a little more accumulated knowledge in the nineteenth century A.D., but possibly the early scribes in Mesopotamia preserved more inherited intuition in the nineteenth century B.C. Being a few millennia closer to early man, could they have been blessed with something we call instinct when it comes to inexplicable wisdom about the animal world? The answer is anybody's guess. The fact is: that modern biologists, following in Darwin's steps, repeat the story that is told with simpler and more poetic words in the early fairy tale of Adam and Eve, who inherited the Garden of Eden. A fairy tale can be told over and over again. The most incredible one is a true one:

THE CREATION THROUGH EVOLUTION

Once upon a time there were no kings, no castles, no people, no forests, no eyes to see and nothing to be seen. It happened so long ago that the time cannot be measured, because time was born in the human brain.

Nevertheless something began to happen. So something must have been somewhere forever, before time began. For if nothing had existed, nothing would have happened. Whatever it was, it was there and set things in rotation. Molecules and solar systems. The intention and power to produce and evolve was there when something fathered the sparkling universe.

On the first day before time began, our Earth was born as a female planet, whirling around in a vacuum with the Moon as future bridesmaid. It was destined to be the one and only chosen planet in the harem of otherwise barren concubines circling around the Sun. Planets are barren when they have no ocean. Planets cannot conceive life without an ocean. Planet Earth became the only one to dance around our Sun with an ocean ready to conceive.

In vain the potent Sun bombarded its harem with fertilizing rays. The face of the virgin Earth was hidden by an impenetrable veil of smoke and damp. No rays, no light came through, and there was no day on Earth yet. But the clock of the timeless universe marked the First Day.

Creativity was active even in the darkness. Atoms danced about and took up positions in orderly alignments like soldiers filling their ranks at an officer's command. Then they danced on as molecules in larger groups. There was omnipotent power, and a plan behind the microuniverse inside every atom. There were possibilities, and there was ability, to fetch raw materials where there were none, or to compose them from nothing. And to provide a quantity of fuel to light up a million billion suns in the universe, while the Earth still rested in darkness under its veil of smoke and fume.

There was still no difference between creation and evolution on the First Day, when the universe was lit. For there was nothing to evolve until something had been created.

The surface of the earth was nothing but a vast, lifeless ocean at the end of the First Day. No continents and no islands broke the smooth skin of the newborn baby. But land was there already, formed as a thin submarine crust of hardened lava. A thin peel, merely enough to shield the outer masses of water from the inner ball of fire where the wild dance of the atoms continued, a dance that was never to end even at the present day. Into this dead ball of water-coated fire was grafted the ability to evolve flesh and blood out of silt and seawater. There were no other raw materials available for evolving man in the beginning. As the early poets described it before the age of telescopes and genetics:

> In the beginning God created the heaven and the earth.
>
> And the earth was without form, and void; and darkness was upon the face of the deep. And the spirit of God moved upon the face of the waters.

And the evening and the morning were the First Day, according to the authors of Genesis. But God was still the only timekeeper. And they tell their readers that one day for God is a thousand years for man and a thousand years for man is a single day for God. Or as Einstein phrased it thousands of years of man later: time is relative. It is a product of the human brain. There were no human brains yet at the end of the First Day. But gravity was there. And gravity saw to it that the water did not run off the Earth, but clung evenly to all sides of the ball like the skin of a new baby.

As the Earth swirled around, the water on top was set in motion. It churned around and around as in a witch's cauldron. There was no land to divert the rotating currents. The moving water was destined to become the blood of future life. It was set in motion to absorb nourishment from the rock and send it in all directions. It was designed as part of a successful perpetual mobile. An impossible invention. As an everlasting automatic heart able to pump water everywhere and in all directions. Able to overcome gravity and send filtered water up into the sky. Just high enough for the precious drops not to be lost into space, but allowed to travel about and fall down again somewhere else, in a vertical circle. Up and down. Up as damp, down as drops. Room was prepared for a future atmosphere. Space was left open for a firmament of air to be formed between the clouds above and the sea below. But there was no air yet. There was nothing but molecules of poisonous gases dancing in the firmament that divided the water running in currents below from the water that drifted in the clouds above. That firmament between the water above and the water below was the only narrow space allotted for later life in the world of our Sun.

In the words of the early scribes:

> And God said, Let there be a firmament in the midst of the waters, and let it divide the waters from the waters.
>
> And God made the firmament, and divided the waters which were under the firmament from the waters which were above the firmament: and it was so.

And the evening and the morning were the Second Day. The drifting clouds were not yet seen by any eye.

Then planet Earth came of age. Wrinkles came as land rose above the smooth sea. Volcanoes swelled up like breasts above water, and entire continents twisted like muscles of hot lava until they slowed down and accepted the spaces allotted to them in the one ocean to be shared by all. Continents were free to float slowly about on the inner ball of fire. And slowly they changed in numbers and shape. But there was forever one single ocean. No matter how big the continents were, they rose up from the bottom of the sea to become islands in a common world ocean.

Man was still not there, and could have been of little help. It took more than manpower to lift continents up from the ocean depth against the force of gravity. It took the power of a planet packed with loaded atoms.

'And God said,' wrote the early scribes of Genesis:

> Let the waters under the heaven be gathered together unto one place, and let the dry land appear; and it was so.
>
> And God called the dry land Earth; and the gathering together of the waters called the Seas; and God saw that it was good.
>
> And God said, Let the earth bring forth grass, the herb yielding seed, and the fruit tree yielding fruit after his kind, whose seed is in itself; upon the earth: and it was so.
>
> And the earth brought forth grass, and herb yielding seed after his kind, and the tree yielding fruit, whose seed was in itself, after his kind: and God saw that it was good.
>
> And the evening and the morning were the third day.

The Fourth Day came. Land and sea were still shrouded in darkness. Heavy clouds sent currents of rain over rocks and waves. The perpetuum mobile was moving alone in the darkness. Water rose to the sky and fell upon sea and land. And the same water rose again to tumble in cascades down the rocks and back into the sea. Running water and breaking surf licked and hammered at the dry rock to soften it, to loosen it up into fine particles of salt and minerals that could be dragged down into the ocean. It was needed there, to form the brew from which the first real life was to emerge. Only plants grew securely fixed to one spot in the darkness. It was hot and damp, a climate best fit for ferns and algae.

But the compact coat of clouds began to split. The rain diminished as the clouds parted and left space for sun and moon to shine through. The virgin earth could see for the first time that the sun was there, waiting with a host of other shining spectators that had been behind the cloak of clouds.

'And God said,' wrote the scribes about the Fourth Day:

Let there be lights in the firmament of the heaven to divide the day from the night; and let them be for signs, and for seasons, and for days, and years:

And let them be for lights in the firmament of the heaven to give light upon the earth: and it was so.

And God made two great lights; the greater light to rule the day, and the lesser light to rule the night: he made the stars also.

And God set them in the firmament of the heaven to give light upon the earth.

And to rule over the day and over the night, and to divide the light from the darkness: and God saw that it was good.

And the evening and the morning were the fourth day.

The Fifth Day came. Now the Sun shone upon the earth. The Sun shone upon land and sea. And the sea was the womb of the Earth, ready with the brew of life. For the first time the Sun did not spill its rays on barren planets. For the first time the rays of the Sun penetrated the salt brew of a virgin womb. And the virgin womb became fertile.

The Sun became the father of all the life that began to move about on our planet. The warm rays of the Sun played all over the surface of land and water in an act of love. The result was the greatest of all miracles in the tale of our planet. The brilliance of the rays woke the slumbering molecules of salts and minerals drifting lifelessly with the currents. They hugged each other in new compositions, and invigorated by the power of the magic rays, they transformed and became the first living cells.

Life began in the ocean. Plant plankton came first. Animal plankton came when there was plant plankton to feed upon. Helpless living cells began to float about, fertilized by the Sun, the first eggs in the long chain of the animal kingdom. There was nothing else from which the evolutionary forces could compose all the marvels that were to follow. Love was to grow from these tiny, innocent single cells. The brutal love of dinosaurs. The gentle love of mankind. Love, and also hate. And all the instincts that chained the animals to given paths and the freedom for man to choose his own trails. In these first minute solar eggs in the ocean lay all that was needed for the evolution of the species and the history of man. They displayed no feelings of love and hate, but in them and from them evolved Cleopatra's passions, Beethoven's Fifth Symphony, Napoleon's wars.

Hunger was born with the first life. Animal plankton and plant plankton drifted together and fed upon each other. The single-celled microanimals ate the single-celled microplants. But when they died and disintegrated, the plant-plankton ate them. As individuals, each living creature died and none survived. But jointly they formed the first simple cogwheels in an ever more complicated perpetuum mobile: eternal life.

By now evolution had been created. The Age of Evolution began with a most successful body building. Cells grew by splitting into two, and more and more joined together to form organs that permitted them all to move, to turn, to see, to open jaws, to swallow and digest, and to dispose of all sewage as food for the microplants. Nothing was wasted. Not a molecule of refuse was left to pollute the ocean.

Creatures of all shapes and with limbs of various kinds began to move about in the sea at will, guided by eyes and other crafty devices that grew as cells were added to cells on the microplankton. Devices of high technology. Molecule combinations that would never have occurred if all the molecules in the world had been tossed into the air and had fallen down to form an accidental pattern. The brain and all its complex nerve system was ingeniously hooked up to even more complicated organs, instinct providing them with direction for use.

All the great inventions for propulsion and communication later duplicated by modern man were installed for the first time in early creatures of flesh and blood. Jet propulsion in the octopus. ASDIC and radar in whales. The eyes of animals developed lenses more perfect and easier to focus than those on any camera. The membranes, bones, and nerves that combined to form the ears of marine mammals preceded any telephone.

On the Fifth Day in the timeless calendar before the age of mankind, the world ocean was teaming with life of all sizes. From the tiny plankton of species not destined to split and grow, to the giant reptiles more fantastic in shape and dimension than the fiercest dragon in science-fiction films.

Some fish and reptiles swam so fast that they shot above the water and could sail like gliders high above the clouds. Living species from the deep sea took to the air long before man learned to walk. Fins of skin turned into reptile wings of leather, and feathers grew out as birds evolved to become acrobats of the air, able to fly and to build nests in trees.

Like the ancestral plankton, all the creatures of the sea and air fed upon each other, yet not even the fittest, with all their teeth and claws and beaks, were able to tilt nature out of balance. The poor and rich lived on together, and sea and air was filled with life but never overpopulated.

As the early scribes put it:

> And God said, Let the waters bring forth abundantly the moving creature that hath life, and fowl that may fly above the earth in the open firmament of heaven.
>
> And God created great whales, and every living creature that moveth, which the waters brought forth abundantly, after their kind, and every winged fowl after his kind: and God saw that it was good.
>
> And God blessed them, saying, Be fruitful, and multiply, and fill the waters in the seas, and let fowl multiply in the earth.
>
> And the evening and the morning were the fifth day.

The Sixth Day arose, with sea and air teaming with living species.

There was still no life on land. But birds and flying reptiles nested ashore. And other reptiles and amphibious beasts crawled up on the beaches and laid their eggs in the warm sand. This was to become the beginning of life ashore.

Out of these eggs came, camouflaged by time, creatures different from those who originally laid the eggs. Instead of fins or flippers for swimming, or wings for flying, they rose up on legs and began to crawl and run around eating grass. Cold-blooded turtles, frogs, and crocodiles were joined on dry land by warm-blooded beasts covered with fur instead of shells or feathers. As if touched by a magic wand, rabbits, mice, and giraffes were among the infinity of new kinds of animals that descended from the reptile eggs abandoned in the sand. There was change and evolution everywhere. But whether antelopes or butterflies, snakes or polar bears, parrots or centipedes, dinosaurs or grasshoppers, they all descended from ancestors that came out of the sea.

There was order in an evolution that could have ended up in chaos. Each species, each biological unit, was so designed that it could only interbreed with its own kind. How then was evolution possible?

The real magic of the Sixth Day is that not only did the wand of evolution pull rabbits out of reptile eggs but that rabbits and crocodiles were pulled out of the same hat, or the same nests. Not all the eggs seemed to have been touched by the same wand, because evolution proceeded along totally different roads. And some of the primeval eggs were not touched by the wand of evolution at all. Cold-blooded reptiles and amphibians still keep on crawling out of their ancestral eggs today. A turtle can creep up on the sand a thousand times, and no hair begins to grow on its carapace. Tadpoles can grow legs and crawl up onto the grass a million times. And yet, when mature, the pregnant females will jump back into the water to lay their eggs just as all toads have done since the ancestors of the species were evolved. Clearly, not everything inside the reptile eggs was tempted by evolution.

Nevertheless, while one species after the other seemed to fall off the ladder of evolution and accept the shape and fate allotted to their kind, others climbed higher. From some nests, or from some eggs in the same nest, beasts crawled forth with hotter blood than that of their cold-blooded, bare-skinned ancestors in the sea. They were in such a rush to improve their own shape that they had no time to wait in the egg but crawled right out of their mothers' wombs, some still so tender and new-made that the mothers needed a pouch on their bellies in which to carry them around until they grew big enough to jump about on their own.

Mammals rose up on four legs and began to breathe, to run, to jump, to climb, and some developed such long fingers that skin grew out as wings between them and they began to fly. Bats descended from mice and other four-legged creatures. And yet other mice, had they tried to jump from trees, could jump until they killed themselves without learning to fly.

Some of the four-legged mammals disliked life on land. The hippopotamus simply jumped into lakes and rivers and retained the limbs they had. But others dived back into the ocean where their ancestors had lived in the days when they were cold-blooded animals swimming about, able to breath in water with the help of gills. They swam so long that they lost their four legs and became whales with a horizontal fishtail. But they would drown if they stopped breathing and tried to develop gills. They would also drown if they opened their mouths and tried to eat plankton or fish underwater, because all land mammals have a passage from the mouth straight into the lungs. The problem for the whale was solved when it got a single blowhole on the top of its head, with a separate tube down into the lungs. But how did the whales survive in the long ages of evolution before the bronchial tubes to the mouth were closed and a separate tube for breathing was led up through the skull to the top of the head? A gradual translocation of the breathing organs over countless generations of diving mammals would be totally impossible.

Yet the making of the whales was a success. The road of

evolution headed straight for its destination and then stopped. It stopped with a perfect product at the end of every branch. And all the species finally evolved, survived and changed no more once they fit the niche in the environment for which they were designed or destined.

Soon there was life everywhere on earth. In every niche, in every nook and corner. Eyes peeped out from every hole. Down from every tree. Up from every pond.

Evolution is the best example of nature's supernatural talent for performing magic.

Endless shelves are filled with literature on evolution. With facts and theories of how it could have happened. The early scribes had the simplest version, condensed to two verses in their poem of the Sixth Day:

> And God said, Let the earth bring forth the living creature after his kind, cattle, and creeping thing, and beast of the earth after his kind: and it was so.
>
> And God made the beast of the earth after his kind, and cattle after their kind, and every thing that creepeth upon the earth after his kind: and God saw that it was good.

And it *was* good. Man was the only species missing at noon on the Sixth Day. The road was paved for his arrival. But Nature's perpetuum mobile, the supernatural clockwork of planet Earth, was in perfect rotation with all its complex cogwheels correctly installed, before man arrived.

Then man arrived. In my childhood home was a lifesize tapestry of Adam and Eve in the Garden of Eden. It was woven by my atheist mother from a special design made for her by a famous Norwegian church decorator. In warm colors it showed a jungle scene with a beautiful couple nude in safe company with a tiger and other jungle beasts. It was stolen one night not long ago, but it remained impressed upon my mind as a good illustration of a topic so often discussed in my home.

Was man evolved or was he created? Or was he created through evolution?

If mankind was molded straight from dust, like the first plankton, why then would he be the last of all creatures to come forth on Earth? Why was the appearance of the most important of all living species delayed to the end of the Sixth Day, when monkeys were already swinging merrily from tree to tree? When gorillas were staggering about in the rain forest beating their hairy breasts as a challenge to anybody fitter than the so far fittest?

The answer is: Man had to come last. All forms of life, the whole ladder up from single-celled plankton to two-legged anthropoids, had to precede him. Creation yes, but through evolution.

But evolution is no less creative than creation. It is creation by degrees over an extensive period of time. Evolution needed a push to get started, and guidance rather than good luck to end up with the production of all living species. It is within human power to dissect and rearrange the elements in the world we have inherited, but it took superhuman power in prehuman time to set the macroworlds in rotation and set the atoms on the road of evolution, from the lifeless molecules of earth and water to the living creatures from which we descend. The discussion of evolution versus creation is a mock fight about manmade words. We cannot dispose of creation by arguing that evolution created itself. Before any animals had the chance to see what mankind looked like, something prehuman, smarter than the apes, must have managed to evolve the human brain.

Science has as yet no name for that invisible natural power behind evolution. The ancient scribes termed it *God*. Scientists commonly scoff at God, because artists generally depict him bearded and in a long gown. Science looks for the formative power as something invisible found here and there and everywhere. Because, everywhere there is life, they find a gene. And science has discovered that inside every microscopic gene is the invisible power that directs evolution and the formation of life. A gene is so incredibly small that there is no room for a bearded man in a long gown inside it. But the total number of genes in the world is infinite, their power is present in everything that grows and moves. In all life, where the ancient thinkers placed the invisible spirit of God, modern scholars detect that there is an invisible force that makes the gene do its duty. If we give priority to the terminology of the early thinkers, *God* was their word for that invisible, ever-present power of command inhabiting and directing all life.

Charles Darwin brought science out of a deadlock last century with his theory of evolution through the survival of the fittest. But no pioneer in science has been more unjustly accused by the clergy and so misquoted by his colleagues. Darwin was not an atheist, nor did he claim that he had proof for his views. He attempted to explain the origin of the species. He launched his view as a theory and was the first to point out the counter arguments.

Darwin's theory was that the species were not independently created, but descended from one another. In the struggle for survival only the form's best fit would survive. This natural selection would cause a natural evolution: the survival of the fittest.

Like everybody else who ponders about the origins of life, Darwin, too, was puzzled by the fact that all giraffes had necks of the same length, and all elephants the same size trunks. What happened to all their ancestors with shorter necks and trunks in the ages of evolution? There were trees with leaves at different levels above the ground, and yet the giraffe's neck and the elephant's trunk had a standard measure independent of the environment. What had happened to the vast and necessarily dominant numbers of transitional forms?

Darwin decided to confer with the fossil findings of the palaentologists. If numerous species had started into life at the same time, he wrote, 'the fact would be fatal to the theory of evolution.'

And proceeding to test his own theory, he actually found that whole groups of species had suddenly appeared together in the lowest known fossiliferous rocks. This discovery made Darwin admit: 'The case at present must remain inexplicable; and may be urged as a valid argument against the views here entertained.'

Again he asked: 'Why then is not every geological formation and every stratum full of such intermediate links? Geology assuredly does not reveal any such finely graduated organic chain; and this, perhaps, is the most obvious and serious objection which can be urged against the theory.'

Charles Darwin saw the problems and never claimed to have solved them. He realized that his theory did not solve all the problems about nature's inventions. To suppose that the eye could have been formed by evolution, Darwin wrote, 'seems, I freely confess, absurd in the highest degree.'

If an eye could not be evolved on a blind beast, then Darwin was the first to realize that something had to be adjusted in his theory of evolution. Someone with imperfect eyes would be a cripple. And a cripple would not survive in the struggle among the fittest.

Modern science has discovered that about a billion years ago a great variety of highly complex sponges, trilobites, and jellyfish suddenly appeared. Some extinct trilobites were then already equipped with more complex and efficient eyes than any of their living relatives today.

The survival of the fittest does not explain the origin of the eye, and certainly not the eye of the earliest trilobites, which died out while those with poorer vision survived. And what came first, the vision or the eye? Without an eye there was no vision. And to develop an eye the concept of vision had to be there.

If man evolved from the family of apes, why then did all the missing links disappear, and not the apes? The apes are there, and man, but the missing links are missing, in the fossils and in flesh and blood. How can evolution have been caused by the survival of the fittest if the apes proved to be better fit for survival than all the missing links between them and man?

Were there jumps in the evolution? Did the ascent go up in steps and not along a smooth ramp? If so, Darwin was on the right road but going at the wrong speed. Evolution goes fast in the womb of a woman. From a fertilized cell through all stages up to a complete human being in nine months. Perhaps there were irregularities in the speed of evolution in the ages when the variety of species were formed. If evolution happened through periodic mutations, in uneven steps, then the fit, the fitter, and the fittest would survive together, each being the fittest to fill the niche in the environment that was vacant for occupation and served to complete the perfect interplay between all species.

Never were the plant plankton more needed than when the birds began to lift from the sea and the beasts crawled ashore. They were the first producers of oxygen. Without their presence in vast quantities there would have been no air. At the time when the first land rose from the sea, there was as yet no atmosphere fit for life. Only a mixture of deadly gases surrounded the planet. Life above water was impossible. But as billions of minute plant cells began to fill the immense surface of the ocean, they produced oxygen in such quantities that it rose from the sea and mixed with the carbon dioxide and other gases to form air. Jointly the living plankton of the vast sea made more than the giant ferns on the green plots on land to pave the road for life ashore.

The discussion of evolution versus creation rests on a question of time. Slow creation is evolution. Fast evolution is creation. If the eye were evolved a billion years ago in the age of the trilobites, it was a fast evolution. Fast evolution is possible, even if time is counted with the measures of man. A tiny egg laid in a pond can quickly become a swimming larva that crawls ashore and clings to a straw while waiting for six legs and four wings to grow out. Then it flies away as a dragonfly. That is natural. Evolution is natural at any speed, as long as we can see how it happens.

Nobody was above the waterline in what the early scribes termed the Sixth Day of God, when the first creatures began to move from sea to land. Gills were exchanged for lungs and oxygen was absorbed by the nose instead of by the neck. The very oxygen was produced by the plankton that was never permitted to grow legs and crawl up on land. They were forever needed in the niches they had been given.

And here lies the real mystery of the true fairy tale of evolution. The fact that not all the reptile eggs laid ashore evolved into rabbits. Not all evolved the same way. Not all evolved at all. Some grew up to be what their parents were, and yet they survived in competition with the fittest.

The magic wand of evolution had touched the eggs with selectivity until all species had their right place in the wilderness, waiting for the arrival of man.

It is needless to ask: what came first, the chicken or the egg. The egg came first. For each generation of every living animal repeats in the egg or in the womb the long evolution since life began. It began with the single-celled egg in the womb of the ocean. There was no new beginning since the first beginning. No interruption of life since life began.

But on the Sixth Day of God's labor, all creativity ended, according to the authors of the oldest known report. Man was the last of the host of living forms to be designed before the day ended. And God used a very honorable model for his final product:

> And God said, Let us make man in our image, after our likeness: and let them have dominion over the fish of the sea, and over the fowl of the air, and over the cattle, and over all the earth, and over every creeping thing that creepeth upon the earth.

> So God created man in his own image, in the image of God created he him; male and female created he them.
>
> And God blessed them, and God said unto them, Be fruitful, and multiply, and replenish the earth, and subdue it: and have domination over the fish of the sea, and over the fowl of the air, and over every living thing that moveth upon the earth.
>
> And God said, Behold, I have given you every herb bearing seed, which is upon the face of all the earth, and every tree, in the which is the fruit of a tree yielding seed; to you it shall be for meat.
>
> And to every beast of the earth, and to every fowl of the air, and to every thing that creepeth upon the earth, wherein there is life, I have given every green herb for meat; and it was so.
>
> And God saw every thing that he had made, and, behold, it was very good. And the evening and the morning were the sixth day.
>
> Thus the heavens and the earth were finished, and all the host of them.
>
> And on the seventh day God ended his work which he had made; and he rested on the seventh day from all his work which he had made.
>
> And God blessed the seventh day, and sanctified it: because that in it he had rested from all his work which God created and made.

The God in the biblical parable ended his creations with a self-portrait. He modeled his own image as a naked couple, man and woman. And on the Seventh Day he rested and was pleased with his work. He blessed that day and sanctified it for man to keep it holy. But the Seventh Day for God was a thousand years for man.

Man was ashamed of his nakedness, and covered the parts God had given him with which to make love. Yet God had depicted himself as a naked man and woman. God was love, explained the early scribes. But man was ashamed of God as well, and painted him as an old man decently covered from neck to feet.

Then mankind went to work. On the Eighth Day they went

to work. To make a better world than the one God had made before he took his rest.

And mankind began to quarrel about the Seventh Day, and formed sects. The Jews said the Seventh Day was a Saturday. The Christians insisted it was Sunday. The Muslims suggested it was Friday. While each of them celebrated a different day to rest and thank God, all of them spent the rest of the week working for a better world than the one God had built for mankind.

And man went out to fight against nature. With ax, with bulldozer, with dynamite. He cut down the Garden of Eden and planted trees in a row.

And man replenished and multiplied. He invented guns and poisons to conquer all the earth for himself. And he subdued all other living things with the intelligence God had given him, and took dominion over the fish of the sea, over the fowl of the air, and over every living thing that moved upon the earth. He killed the big beasts and the little ones and he built cities for his own kind. And if his own kind built other cities and cultivated other lands, he made war upon his own kind in other cities and in other lands.

And man worked on the Eighth Day while God still rested. And one day for God was a thousand years for man. And man worked for thousands of years while some took Fridays off, some Saturdays, and some Sundays. And some took no days off at all, and trees and animals fell at the touch of the tools and arms of man every day, for millennia upon millennia during the Eighth Day of God.

Trees were stuck to the ground and could not escape. Animals had fins and wings and legs and could escape faster than man, but not as fast as man's bullets, nor were they smart enough to escape man's nets and traps. Man needed the blubber of the whales to produce dynamite, and more fish than rich people could eat, to get richer. Then money was put into the bank and into the arms race. And man killed his own horses, his own donkeys, and his own camels when he built himself cars and tanks.

But in the eleventh hour of the Eighth Day, before the sun had set and while God still kept his temper, man saw the maltreated vestiges of nature left outside the city walls and along the asphalt roads. And man got curious about how things could move that had not been made by man. Man could make plastic trees, but the trees in the forest moved by themselves out of the mud. Man could fill automobiles with gasoline and make them move, but beasts in the forest filled their own bellies with water and grass and ran where no wheels could run. Man could make airplanes with stiff wings that needed long runways and costly fuel to lift and land, but a simple fly could fuel itself with dirt or dung and take off from the point of a pin in any direction, and land upside down on the ceiling. Was there an inventor smarter than man? Yes. But where?

So man began to look into the things in nature that still moved by themselves. He found they were pieced together from parts so small that they could not be detected by his own eyes. To see what nature had put together, man made himself magnifying lenses like those in the eyes of fish, and telescopes like those in the eyes of eagles. And to detect what the eyes could not register he built himself antennae like those on the heads of bees and bugs, and sounding devices like those in the heads of bats and whales. And man thought that with all his instruments he could observe and measure all there was. And that what he could not find with all his instruments, that was not. And none of man's instruments helped him see God.

So man was sure he was alone in a world of animals and trees, and the only one to design and invent.

And man decided to create men in his own likeness. He built robots of metal and put computers into their heads. He built his robots in the image of an armored soldier, men with breastplates of steel and jointed limbs, men without women and with no sex or ability to love. On the Seventh Day, while God rested, man created men to his own taste from plastics and metals, men that could not think but obeyed orders to stagger around for entertainment on the floor or to man a tank in combat.

But man grieved at his own inability to give his robot a brain. And he marveled at nature that had evolved man's perfect brain out of the brain of a monkey.

And man studied his own body and his own brain and was impressed. And scientists and sages asked each other if there was a God without a brain who was even smarter than man. But though they used telescopes and spacecraft, scientists found no God on the moon or elsewhere in the universe, and no angels in the clouds. So they began to look for the power of evolution inside the living cells, and they found the genes. They found that the genes were incredibly small and yet loaded with the magic power of deciding how every species should grow and take shape. How every leaf and flower should form on a tree. How every tooth and eye should be shaped on a beast. Each microscopic gene seemed loaded with divine power, and there were genes in every living cell on earth. But no visible divinity. No God was seen either in the greatness of the universe or in the smallness of the genes.

But the eyes of man were formed by the power of the gene. And eyes were made by the gene in such a way that they could only look out of the head. No man can see into his own brain or into his own heart. Love comes from somewhere within. And if it is true that God is love, then he is hidden on the blind side of all eyes.

In dissecting life into its most minute component particles, modern scientists have found themselves trapped into a totally impossible deadlock. They have found something termed *protein*, which is absolutely necessary for the formation of any kind of life. And they have also found something else which they have termed *DNA*, which in turn is absolutely necessary for the formation of proteins. But proteins are also absolutely necessary for the formation of DNA. Neither of the two can be formed before the other is there already. Scientifically speaking, life can never have evolved unless DNA and protein had been created simultaneously and independently.

But science has discovered that it is mathematically impossible that either of the two evolved by chance. DNA is composed of five parts, which have been termed *histones*. The likelihood that even the simplest of these histones are formed by chance, is estimated to be one in 20^{100}, a number larger than the total of all the atoms in the visible part of the universe. The molecules of the proteins are also exceedingly complex. The chance that the first proteins in the ocean are formed at random from drifting atoms has been calculated by scholars to be one in ten followed by 113 zeros, a chance dismissed by mathematicians as never happening.

At the end of the labyrinth of the microworld, science has thus found the mathematical proof for the beginning of life. Neither DNA nor protein can have been formed by chance. Both must have been created. And both must have been created at the same time, since one depends on the other for its existence. Thus life as such began through creation.

Science has found terms for all created things, but it still lacks a scientific term for the creator. There is a vast number of terms to choose from. Among those first put into writing was the one used in Genesis.

While some learned men took life apart to study its tiniest particles, others put the tiny parts together into combinations never attempted by nature. Man wanted materials that did not decompose. And he wanted poisons to kill the bugs and microorganisms that tried to stop him from starting monoculture or filling the planet with his own breed.

Man put atoms together into everlasting poisons to fight weeds and bugs. The inventor of DDT was awarded the Nobel Prize. A few years later, DDT was forbidden in all developed countries and those who used it were punished by law. But developing countries had become dependent, and continue to spray DDT on pests that build up resistance. DDT and thousands of other non-degradable chemicals with molecule combinations designed to kill, continue to kill forever, wherever they blow or float in our own environment, into the ocean, or into the soil.

Man was after all not smarter than nature. It was no chance but clear intention that made nature build up only molecules of the kind that could come apart into atoms that would group together again into other forms of life in a never-ending rotation of reincarnations.

XVII

Progress from Paradise

THE OCEAN WAS CLEAR when we sailed on the *Kon-Tiki* raft in 1947. We were in intimate contact with the plankton that we scooped up with our hands. And even in intimate contact with the distant galaxies, our only companions above sea level apart from the sun and the moon.

When preparing to confront the world ocean a second time, by reed ship in 1969, I assured my companions that our greatest pleasure for the coming months would be to live eye to eye with the marine world and see the marvelous purity of the sea water. We had drifted 8,000 kilometers across the Pacific on our balsa-log raft, without seeing the slightest trace of pollution. No evidence whatsoever that there were other people than us on this planet.

The second time, two decades later, we were to attempt a

crossing of the Atlantic. On an African type of raft-ship built from bundles of papyrus reeds. The very first morning we woke up off the coast of North Africa, and found ourselves sailing in a soup of glittering oil and asphalt lumps. Our reed ship, *Ra*, carried seven men from seven nations. Abdullah, our Muslim from Central Africa, could not wash his face in honor of Allah because of the filth. We sailed out of the soup, and I told my men that we had just happened to pass the wake of an oil tanker flushing its tanks before entering port. But to our surprise, as we sailed away from Africa, no matter how clear the water was, almost every day we sailed past solid clots of asphalt, varying in size from rice to potatoes.

We sailed under the United Nations' flag to show that men of different skin color, religious faith, and political upbringing could collaborate in peace even under stress and in cramped quarters. Shocked at the oil clots and drifting plastic bottles we observed, I sent the first radio message to the United Nations secretary general reporting our observations: the world ocean was getting visibly polluted.

The ropes holding our papyrus bundles together chafed off and we lost most of the reeds in the starboard half of our ship shortly before reaching Barbados. So we abandoned the floating half of *Ra* and determined to try again with a better built reed vessel.

In 1970 we sailed our second papyrus ship, *Ra II*, again under the United Nations' flag, with the same crew augmented to eight. This time Secretary General U-Thant asked us to make a day-by-day record of what we saw, and to collect samples. Madani from Morocco was left in charge. Equipped with a dip net and a logbook, he began to fish up oil lumps. Tiny and large tar balls. *Ra II* crossed the widest span of the Atlantic from Morocco to Barbados. We picked up oil clots on forty-three out of the fifty-seven days the voyage lasted.

Our samples from *Ra II* were delivered to the Norwegian

delegation of the United Nations and examined by experts. It was found that the samples came neither from leakage on the ocean bottom nor from the wreck of a single tanker. We had sailed through the combined spillings from the world's vast fleet of modern supertankers.

Our report was published as an afterword to the secretary general's own report on the State of the World Ocean for the first Law of the Sea conference held in Stockholm. For years I was to become more involved in concerns for man's future than for man's past. While professional oceanographers sent research vessels into the mid-Atlantic to check, and came back to report that the pollution was even worse than our warning, I felt impelled to travel all over the world to testify to governments and scientific institutions on what we had seen and what it implied.

There was only one world ocean and it could not be divided between nations. The ocean bottom, yes. The rock bottom was always there, but not the living sea. We had experience. Step onto some floating bundles of reeds in the national waters of Morocco, and you can step off in the national territory of Barbados a few weeks later, drifting a bit faster than the pollution. But all the oil clots follow in your wake. What are African waters in one season are American the next. Hoist a red-and-white Peruvian flag as you board some floating logs in Peruvian national waters, and you will have to exchange it for a French tricolor when the same waters enter Eastern Polynesia. There is only one world ocean, and it churns around to brew the soup of life. Ever fresh, as in the remote ages when the first protein and the first DNA were created together to form life.

The first to react before any governments took any action were the owners of the tank fleets. It was prohibited to dump sludge from oil tankers into any part of the ocean. The shipping world was living off the sea and willingly imposed restrictions. But the oil clots had been nothing but visible eye-openers. What the scattered ships disposed of far apart was a mere drop compared to the constant flow of pollutants from all the cities and all the farmlands on all continents. There is no river in the modern world that enters the ocean without seepage from sewer systems or fields polluted by spray. DDT and other chlorinated hydrocarbons exist in whales swimming in the Arctic. No matter how tall we build our chimneys or how far we bring our pipelines into the sea, the nondegradable pollution we like to dispose of is forever with us. Nothing blows off the thinly coated roof of our atmosphere, and nothing runs off the edge of the planet beyond the horizon.

The ocean seems endless, except to raft voyagers and astronauts. But it merely winds around the continents like a lake, with land enclosed inside and on all sides. And it clings to the planet like the peel of an orange that also seems endless, because it begins again wherever it ends.

The ocean has lived for an estimated four billion years, to judge from bacteria and other microorganisms imbedded in the oldest rocks. Throughout all those ages, the ocean has been the global filter. Untold masses of pollution from carcasses, excrement, and rotting vegetation have entered the one and only ocean, from all the rivers or with silt from all shorelines. The number of prehistoric monsters, the whales and fish and plankton that have died and decomposed in the sea would have been enough to displace all the water if not decomposed and recycled to new, young, smaller life. The molecules of all matter entering the sea from land and air decompose and are rebuilt, and nothing but pure water is allowed to evaporate and rise to the clouds. Thus the clockwork has moved smoothly and in the interest of global life and sanitation for billennia, the sea sending clean water back to the land with the clouds, and gravity sweeping all dirt downhill into the sea, where it is digested and transformed. The clockwork was built as a perpetuum mobile, and as such it would work forever with the ocean as a filter, if man had kept his modern molecules tied up in his own laboratories.

But after one-time use, man no longer wants his everlasting molecules, nor any control as to where they all end. Some pile up on city dumps, on beaches, and along roads as harmless but ugly plastic scrap. But most detergents and insecticides are absorbed as by blotting paper into all organic life. And once inside they can never again get out. They enter the life cycle of plants and are eaten by beasts. They follow the universal current of the life cycle, which ends up in the sea. But while all the molecules of nature are eaten by microorganisms and are digested to form new living cells, the manmade ones are indigestible, nondegradable, and enter to be stored in the living cell. Most of the hydrocarbons float and stay in the surface layer. And that is where the plankton lives. And most of it comes from outlets along the coasts. And that is the breeding place and home of the bulk of marine life.

A constant stream of sea water runs through the body of plankton. Nature's molecules are digested. But not those composed by man. They get stuck in the body of the plankton, and plankton is drifting everywhere to act jointly as the ocean's giant vacuum cleaner. They leave the water clean but are themselves swallowed in vast numbers by mollusks, crustaceans, and fish. Thus the manmade molecules accumulate in ever stronger concentrations as they follow the food chain up to man's cooking pot. And no matter how much we cook and how much we chew, we can never destroy the nondegradable and venomous molecules we have produced in an unfortunate attempt to improve our environment.

What the farmers and the housewives spray out of plastic bottles, the fishermen and the middlemen serve us on our own plates.

My childhood fear of the ocean had left me on the balsa raft. My fear was now instead that man should destroy the ocean. A dead ocean meant a dead planet. Neither God's evolution nor Darwin's theory would help us fill our lungs with air if we killed the plankton and stopped the planet's oldest production of oxygen. Only 30 percent of our planet is dry land, and a fraction of that land has forest vegetation. The plant plankton is today our main source of oxygen. The forests are shrivelling on all continents, falling to man's fires and bulldozers. On the survival of the mini-flora of the sea depends the survival of mankind and all lung-breathing species.

From the plankton, the oxygen rises into the air and the winds spread it over sea and land. Winds and marine currents travel together from east to west in the tropical belt of the oceans on either side of America. The voyages of *Kon-Tiki* and *Ra II* across the Pacific and the Atlantic oceans had been pure drift voyages. Left alone, any buoyant vessel will float westward from Africa to America and from America to Polynesia. No navigation is needed. The rotation of the earth keeps sea and air in constant westward drift in these areas. Not so in the Indian Ocean. Strong seasonal variation in the climatic zones of Asia and Africa produce the monsoon winds, which blow in opposite direction summer and winter, forcing the currents to change as well. This was a challenge to us, who by now began to ask ourselves: Could the reed ships of ancient times actually have served for intentional navigation?

This prompted me to build the reed ship *Tigris* in Iraq in 1977 and set sail into the Indian Ocean. Little did I suspect then that this, my fourth experiment with a prehistoric type of craft, would give me warnings of still another imminent peril for the future of our planet. Climatic disturbances. In Sumerian style, we built our test vessel this time from local *berdi* reeds. We were eleven men from different nations when we embarked and sailed down the rivers of former Mesopotamia, from the conjunction of the Euphrates and the Tigris, where Arabs, Jews, and Christians alike place the legendary Garden of Eden.

We had chosen to sail in the winter season, a time when both meteorologists and fishermen had guaranteed us the north wind at our back, out of the Persian Gulf, and then the northeast monsoon to push us from Asia to Africa. We sailed for five months with tons of food and water stored on board, but not for a single day did we get the wind direction we had expected. The north wind failed in the Gulf. Instead, a contrary south wind threatened to throw us ashore in Kuwait as soon as we came out of the river mouth in Iraq. We were forced to pay ransom to pirates from Failaka island to tow us away from their reefs before we learned to sail our reed ship into the contrary winds. When we sailed through the Hormuz Straits and entered the Indian Ocean, even the winter monsoon failed us completely. The fishermen all along the Persian Gulf and on the coast of the Indus Valley assured us that the prevailing winds had started to blow in unpredictable directions the year before, in 1977. As we gained experience we managed to sail close-hold from the coast of Oman up to Pakistan, and with no help from the monsoon we fought the changing elements at the cost of a broken topmast to cross the Indian Ocean from the Indus Valley in Asia to Djibouti in Africa at the entrance to the Red Sea.

Five months on a raft-ship had let us feel on our bare skins what the fishermen along our route had told us. The seasonal winds, on which local navigators had based their itineraries for untold generations, had begun to bring confusion into nature's age-old clockwork in the Indian Ocean. For the first time in human history the weather gods had lost control of their ingenious climatic transport system, a system known to man since long before meteorological records began.

We from the western world celebrate Vasco de Gama as the discoverer of the sea route to India, because he learned the secrets about the monsoon winds from a medieval Arab. We tend to ignore the fact that the Indian Ocean was a highway for maritime traffic long before medieval times, and that the direction of the monsoon winds were already a matter of record by then. In the first century after Christ, the Roman historian Pliny the Elder wrote about the truly astonishing amount of trade that was carried out in his time by ships between Egypt and ports in Ceylon and China. He recorded that the trade route from Egyptian ports on the Red Sea across the Indian Ocean went back to the days before the use of wooden ships, when the early Egyptians navigated 'with vessels constructed of reeds and with rigging as used on the Nile.' He stated that the Egyptians began their voyages from the Red Sea ports in midsummer, at the time the dogstar rises. Further: 'Travellers set sail from India on the return voyage at the beginning of the Egyptian month Tybis, which is our December, or at all events before the sixth day of the Egyptian Mechir, which works out at before January 13th in our calendar.'

Thus the Romans two thousand years ago had learned from the early Egyptians that the monsoon blew with such dependable precision that mariners set their sailing departures according to the days in the month when the wind would change.

Perhaps I might have thought no more about the shocking absence of the monsoon the year we crossed the Indian Ocean, had it not been for a very effective reminder. After all, some meteorologists tell us that it is normal to have anomalies. Others, however, are worried, and observe that we suddenly get hurricanes twice or more in a year in parts of the ocean where there were never hurricanes before. Whatever be the case, we are living in an age of human abuse of the environment, so we ought to be on the alert. And I was indeed alerted soon after our landing among the starving crowds of Africa. A couple of years later, curiosity lured me back into the ocean we had struggled our way across with the reed ship *Tigris*. Again I was driven by my interest in man's earliest conquest of the world ocean.

It started like a detective story. One day I got a letter from a stranger with the photograph of a huge stone head of a long-eared giant buried to the chest in sand. What was the mystery? The statue had been discovered on one of the twelve hundred coral atolls in the Republic of Maldives. A tiny Muslim island republic far from anywhere out in the Indian Ocean southwest of Sri Lanka, alias Ceylon. The early Egyptians must have passed here on their way to Ceylon and China. The sailing route south of India was blocked by the extensive mid-ocean barriers of the Maldive reefs, except for a passage at the equator. The Maldives had been a rigidly Muslim archipelago ruled by sultans ever since it had been converted by the Arabs in A.D. 1153. Muslims do not allow any representation of human figures. So the long-eared giant must have been left by people who had found the Maldives before the Arabs came.

At the invitation of the Maldive president, and with a team of archaeologists from Norway, I returned to the Indian Ocean by air to find that the mysterious limestone bust had been pounded to pieces by fanatic Muslim islanders, except for the long-eared head, which was saved. It was part of a most beautiful Buddha.

Archaeologists had never been tempted to explore this oceanic archipelago, because of the old dogma that old-time seafarers hugged only the continental coasts. We were thus the first to discover that the Maldives were full of prehistoric remains. In the forests, sometimes visible from the sea, were large pyramidal structures of limestone blocks and coral rubble. They had once been beautifully dressed by carved slabs later stolen by the Arabs to build their twelfth-century mosques. These islands had obviously been an important stopping-off place for early voyagers between the East and West. A convenient center for merchant mariners from the Middle East, the Mediterranean world, India, and China ever since ocean navigation began and civilization started to spread. There were images of Buddha and Hindu deities in both stone and bronze. Lion heads, reliefs of oxen, and lotus flowers, phallic monuments and stone masks of elegant men with slender, curved mustaches such as were typical of ancient Yemen.

Firm traditions in all parts of the Maldive archipelago insisted that the large pyramidal mounds were built by the Redin people, a people of great navigational skills who had been the first to discover and settle their islands. A few years later I was allowed to enter the then closed People's Republic of South Yemen, where I recognized the stone masks of antiquity that were familiar from the Maldives but were of neither Buddhist nor Hindu origin. And Russian archaeologists, excavating in the former Yemenite kingdom of Saba, had recently found hoards of the shells the people of the Maldives used as money. But what clinched the evidence were old inscriptions preserved in Yemen. They frequently refer to the Reidan people, who ruled the entrance to the Red Sea and the adjacent coast of South Arabia in early times. It was the Reidan merchant mariners who caused the queen of Saba's wealthy empire to collapse when they found the short cut to India by sea and thus made the lucrative caravan route through Saba's mountain kingdom redundant. Nobody cared to pay tolls for passage through Yemen and still have to drive their camels all around the Persian Gulf to bring cargo to and from the lands on the other side of the Indian Ocean when they could load everything on a boat and have it carried directly to India and back by merely sitting with a steering oar.

It made sense that the Reidans of South Yemen were the Redins who came to the Maldives and left their huge mounds and self-portraits of elegant men with mustaches. They had reached these oceanic islands before the Hindus and Buddhists of continental Asia and Ceylon took over, and ruled the archipelago until A.D. 1153, when the Arabs came with the Koran.

Sculptures and ceramic shards took us far back through the forgotten ages of Maldive prehistory. But once again our efforts to penetrate into the unknown past of man were brusquely interrupted by warnings about our equally unknown future. We had come to the Maldives in the dry season to make archaeological work possible. But suddenly a deluge of rain poured down and filled our trenches. We were excavating the well-preserved base of a solar-oriented square pyramid with a ramp approach and sun symbols in high relief, and had to hurry to erect poles supporting canvas to protect everything exposed. Contrary to all assurance from meteorological registers and Maldive people, the rain poured down in torrents, soaking us to the skin and filling our jungle boots until we emptied them as if they were pitchers.

The monsoon had lost its way. The heavy cloud carpet tipped off its load in the wrong place. The rain splashing down upon us on these atolls in mid-ocean came from the damp jungles of tropical Asia and was destined for arid zones in Africa. West of us lay Ethiopia, expecting it and needing it, but reporting disastrous drought. Nature's climatic metronome seemed off the beat. In meteorology it was becoming normal to have anomalies. For modern man it was no longer safe to trust the old Egyptian sailing directions of steering from Africa to Asia when the dogstar rises and home again before the sixth day of the Egyptian month of Mechir.

Everybody talks about the weather, because the weather changes. It has always changed. The global climate varies too, but over the course of millennia. To the man in the street and the man in the field, weather comes and goes with the clouds, and the clouds come with the wind. The weather is sent by God, and no windbreaker or umbrella made by man is big enough to change it. The degree of humidity and direction of wind – that is the weather.

Cut down the forest, and the humidity disappears. Turn the jungle into a desert and the winds change. That is what we do. And men of all nations have done it systematically and effectively over a very long period. Together we are fiddling with the bags of the weather gods, but we do not want to take the blame for the result. If the climate is changing, we need more study to be sure it is our fault. We need more time to cut down more rain forests and send more smoke and spray into the sky until we have proof that we are able to upset the balance in nature and derange the weather system. Nature itself remains the suspect until environmentalists can prove that the climatic anomalies are not merely normal freaks of nature. The burden of proof rests on those who shout warnings and accuse man.

Certainly, nature's climatic clockwork has its irregularities, and many are neither caused nor understood by man. As I write, I wake up from my memories of the splashing cloudbursts in the Maldives, and through the window I see the bone-dry, sun-baked pyramids of Tucume. Over there, under a spotless blue desert sky, our archaeologists are digging drainage trenches and erecting roofs over our excavations in preparation for a real freak of nature. An expected Niño year. Seven years have passed since the Niño Current hit the north coast of Peru full force in 1983, when coffins from the cemetery drifted like canoes in the main street of Tucume. The disaster usually comes at seven-year intervals and we have to be prepared. The deluge from the Andes and from the sea was exceptionally violent last time and caused incredible damage all along the coast. But our excavations show that the Niño Current has brought periodic disasters and deposits of mud over this desert landscape since long before modern technology could begin affecting the weather.

In front of me lies a pile of papers and clippings about a different threat to our own environment. Warnings from United Nations Environment Program, World Wildlife Fund, Friends of the Earth, The Sierra Club, The Club of Rome, and scientific academies in many countries. A fresh clipping from a leading Norwegian newspaper has a report from the Arctic Sea under the following three headlines:

EXTENSIVE ICE-MELTING IN THE BARENTS SEA.

THE ICE MASSES IN THE BARENTS SEA HAVE DIMINISHED BY 25 PERCENT IN THE SUMMER MONTHS OVER THE LAST 20 YEARS.

SCIENTISTS AT THE POLAR INSTITUTE ASK THEMSELVES IF THIS MAY BE DUE TO THE GREENHOUSE EFFECT.

It's scary, says the average adult at the breakfast table, and puts down the newspaper. And goes to get his car. And every car on this planet continues to spew exhaust, while heavy industry and supersonic flights join them in sending harmful gases into the atmosphere in competition with the healthy oxygen from the plankton of the sea. And scientists measure and test. And agree that the percentage of carbon dioxide has increased in the atmosphere since the beginning of the industrial revolution. Millions of tons of fuel are sucked from the soil and sent up as smoke and fumes into the air, every hour around the clock. This pollution will stop the sun's rays from reaching us, and create a colder global climate, predict some meteorologists. No, say the others: it will form a ceiling retaining the heat lost from the earth. And while the scientists quarrel and disagree, the ice melts in the Arctic Sea and glaciers run away. Is it getting warmer? If so, we cannot prove that the change is due to any act of man. Umbrella effect or greenhouse effect, it can still be a freak of nature.

But people are getting increasingly worried everywhere. Especially people living on low islands and lands at or below sea level. The Maldive islanders, on their beautiful coral atolls six feet above sea level, are getting alarmed. The coral reefs have protected their huts, their garden plots, and their coconut groves around calm lagoons since the days of the Egyptian reed ships. Now they have been told to be prepared: their islands will submerge if the glaciers of Greenland and the Antarctic melt.

Looking at the row of pyramids outside my window, I see a fragment of the human past. This world has seen a great many civilizations. And many of them have survived for longer periods than ours up to the present. They were all as sure as we are today of having founded the first eternal civilization. We today differ from them in having our western civilization spread to embrace the entire planet, leaving no room on any continent for any other culture to take over if we fail.

We are today erasing our past and transforming the present of all other cultures, convinced we are on the right road. But no one, absolutely no one, can tell us what the final appearance of our rapidly changing technological civilization will be. It has not yet been planned. It cannot be wholly foretold.

While nobody is planning our global future, there is no lack of prophets. We are guessing at the result we are ending up with. Only future generations will know. A generation ago we blessed every invention that was made, even the atom bomb. But today, at a time when we plunge into the technological era with fairy-tale visions of a manmade environment, science itself begins to see that nature is totally superior to man in its incredible composition of the world's ecosystem. Destroy it, and no brain and no money in the world can put it together again. Today, when we are on the verge of an irreparable catastrophe, an army of environmentalists has grown up almost overnight in all civilized nations. Scientists and laymen alike. Mankind is at a dramatic crossroad, and the prophets are divided in their predictions. Optimists rejoice in the conviction that man will soon move up onto comfortable space platforms, so never mind what we do to planet Earth. Pessimists predict that the arms race will end with a nuclear holocaust that will send us into space, piecemeal and without spacecraft. But let us trust that there is still time for better solutions. Today we see the reefs and can grab the rudder and steer for happier destinations. If we really believe we can build ourselves a better world than the one that made mankind see the daylight, the one that helped us breathe and eat on the Seventh Day, then let us aim for it. But we must know where we want to end up. How far we want to squeeze ourselves into a press-button world of machinery, robots, and physical inactivity. What is our final goal? Do we know?

At the moment the dreams of a life in outer space have a magic grip on people the world over. With good reason we admire the brainwork that has made it possible to shoot courageous human beings like bullets into space. Allowed them to circle weightless around the earth like autumn leaves in the wind. And to land for a walk in dust and gravel on the moon. Human curiosity cannot be restricted by geographical boundaries. With telescopes and spacecraft we will know ever more about other worlds in the universe. We already know enough to ascertain that, within our own solar system, there is no better place to land than on our own planet. We know the moon is a desert more barren and lifeless than the Sahara. We know that Venus is a flaming inferno as hot as our image of hell. Maybe we can pick up some useful minerals from the frozen wasteland on Mars, and probably those who can afford it may get excitement out of a tourist flight to the moon.

But our legitimate fascination for outer space must not lead us to teach the younger generation that our own planet has less to offer them than its uninhabitable companions circling around the same sun. We must not let our technological successes sweep us completely off the ground. That seems to happen, however. We entertain our children and ourselves with books and films on imaginary space travels to marvelous planets. We amuse them with space toys and with comic strips of horrible interplanetary wars. And all over the world there are individuals and clubs dreaming of and speaking about extraterrestrial visitors. Some believe they see green little men coming out of flying saucers. Books sell by the millions the world over if they argue that people of early Egypt, Peru, and Easter Island were ignorant barbarians and that therefore the pyramids, the Nazca lines, and the stone colossi on Easter Island had to be the product of extraterrestrial visitors. If supermen from outer space come on visits in flying saucers today, they appear to be so horrified at what they see of our modern communities that they rush back again to the nearest star after a few seconds' visit. They have a long voyage home, for they come from other solar systems, and the nearest one is a thousand years away if the saucer is able to travel with the speed of light. We permit our youth to abandon reality and fill our own fabulous home with planetary escape dreams.

The extraterrestrial visitors obviously do not like it here. But the true-to-life terrestrial astronauts, American and Russian, come back to Earth and spare no superlatives in describing what a jewel of a planet we live on, and how marvelous it is to come out of the capsule again, remove the helmet, and inhale the smell of green grass. I met Neil Armstrong when he returned from the moon and joined us in a World Wildlife Fund drive as a devout activist in global environment protection. I also met Oleg Atkov the day he came down to Earth after a record flight of eight months as medical doctor in outer space. He staggered from the space capsule, a former atheist suddenly filled with humility and respect for the creative force that had composed the fertile soil on our planet and the seeds it could transform to the benefit of man. He was sent to the Caucasus for rest and time to pull himself together, for the bones in his body had separated a couple of centimeters due to the long months of weightlessness.

In spite of all the lessons we have received from astronomers and astronauts, there is a growing number of space dreamers who keep on telling the youth of our planet that man's future is above the clouds. They tell us not to worry, for if we mess things up too badly here on earth, we can always move up onto manmade platforms in outer space.

But can we?

With the accelerating speed of technological progress some wealthy superpower or oil-rich nation with money to spare will one day be able to send up parts to build large platforms in space.

A kind of miniplanet, like the oil-drilling rigs in the ocean.

With sterile soil and plastic grass and flowers. But with much of the inhabitable land on earth already crowded, how can the space engineers help all of us to get room on flying platforms? If they can lift up the citizens of a single city, or people from a single little island, what about the billions of human beings who have to remain behind? Is their driving force the fascinating technological challenge of building space platforms, or an interest in the well-being of future generations?

What kind of future can be offered the select group of people we could afford to shoot into space? These privileged emigrants would soon come down to look for what's missing. To stretch their legs and pull their bones together. To take off their oxygen masks and draw in a breath of unmanipulated air. To ask the farmers on the ground for worms and bacteria to make their sterile soil fertile. To catch insects needed to fertilize their plants. To hoist more water up to their desert platforms, which are on the wrong side of the clouds and not blessed by rain. To fill their tanks so they can breathe in outerspace, as no oxygen comes from plastic plants. To fetch more building material so they can tow a forest-covered platform behind their own. In short, they will find it necessary to bring up into space what man ignored in his daily life on earth.

Those who would get their space dreams fulfilled would awake to reality and understand why the environment the evolutionary forces gave to man is not made from plastic and chrome. They will regret the money and efforts they wasted to be propelled in voluntary exile into outer space. Their experience would be the best lesson in the ingenious buildup of the ecosystem and the need for environmental protection. They will feel gratitude for the marvelous green and fertile space-ball on which we all are traveling together, at no cost, and with no maintenance problems. Nothing is better than a genuine experiment to prove or disprove whether something will work that has never been attempted. Perhaps the refugees above the clouds will return with their feet truly on the ground and become the new prophets we really need to waken a slumbering humanity.

Man has conceived a bedevilling word: 'Progress.' First we let it out of our mouths, and next we permit it to pull our noses and lead us astray. What is progress? The word was conceived to describe a forward motion or an improvement in quality. Then, with little modesty, we took it for granted that anything we produced ourselves was better than what nature had evolved. We began to consider nature as an enemy that had to be controlled. Any weapon against nature, any invention that would take us another step away from our own shameful past in the wilderness, was considered a move in the right direction. Looking back through time, man assumed progress could be measured by the clock, by generations, by centuries.

Progress can be measured horizontally on a map but not vertically on a calendar. The level of culture was higher in the Middle East 5,000 years ago than it was 500 years ago in medieval times in Europe and the Middle East.

Thanks to medicine, the arts, education, and communications, we have produced great advantages for the minority of mankind that can afford them. Our only fear is that overconsumption may set a limit to growth and to how long our present living standard will last. We speak of the possibility of a future catastrophe. But those of us who speak in public or write forget that the majority of mankind today are already hit by that catastrophe and still exist in living conditions as miserable as those of the Middle Ages.

We who have enough want rich and poor to have more. We have mercy for laborers who work with their bodies, and so we invent push-button systems and computers to relieve them of physical toil. And then we send them unemployed into the streets, while those of us who can afford it exercise by pedaling stationary bikes.

Our 'progress' has become in part an effort to complicate simplicity. The farmers and fishermen remain the only true nobility of modern society, working to feed us from the natural world we left behind. Without them, modern society, with all its banks and shops and powerlines and water pipes, would collapse.

A global surplus of money allowed us to embark on an armaments race. We have killed almost all the lions and tigers and agreed to protect the few left, meanwhile concentrating on sales of weapons designed to kill our own kind. If enemies are far away, we store up astronomical quantities of long-range rockets and super-superbombs, and then we scrap them to make better ones. Modern ideas of progress produce holocaust armaments better suited for killing more of our fellow humans at longer range.

Who is the enemy in the civilized world? The enemy changes overnight, and all of us are so alike that we have had to put on uniforms to enable us to see the difference. But at the new, longer ranges we need no uniforms; anyone elected by the people or by himself can sit and press a button to end the armaments race. Self-defense has progressed to self-destruction.

We thought progress meant better weapons with which to defend ourselves against hostile neighbors, so we might enforce peace on our own terms. But we have learned from experience that whether we invent bows and arrows or nuclear bombs, the enemy copies our weapons and turns them against us. We have built machinery to free the laborers from backbreaking work. But the machinery took over the labor and left the spirit-broken laborers idle.

As the progress in armaments has surpassed the limits of insanity, man has stirred up an enemy we all share. And history has shown that alliances between nations are created by the sharing of a common enemy rather than a common friend. As we begin to detect this threat from over our heads and under our feet, we may come to our senses, scrap anything designed for manslaughter, and join hands in defense of people of all races and nations. Camouflaged by the foliage of progress, in the soil and the clouds, our enemy is released from our spray bottles, exhaust pipes, sewer systems, and industrial chimneys. We have no catchy name for this enemy as yet. Environmentalists of the present century began to use the term 'pollution,' the most negative aspect of technological 'progress.'

It is our own environment that is in slow revolt against us – the friendly ecosystem that bred us and fed us for millions of years. Now, its law of equilibrium in a *perpetuum mobile*, it is attacking us with our own nondegradable molecules, and our overproduction of carbondioxide. Today, with our partial understanding of the atom, we are so many and so powerful that we are able to cut off the top branch of the biological tree of which we are a part. Enough of the trunk and root may remain to sprout anew. Our branch was once the topmost one, and may be removed, but the tree will remain standing as it did before man appeared. Crippled, perhaps. But man's future will always have to be at the top of a living tree, its sap drawn up from roots firmly set in the magic of soil.

If there were to be a testament to my own children, I would say:

You are now to take over this planet; take good care of it. We did not, when we borrowed it before you. It has been a wonderful playground for us and for those who lived before us. But we learned from our parents to look for progress, and we misinterpreted the word. Our parents, and their parents since the time of Abraham, have told us of a God who gave mankind a paradise with a lovable naked pair to represent him. But none of us understood the parable of Adam and Eve in the Garden, which was planted and populated to God's satisfaction before the morning of the Seventh Day.

Forgive us for the forests we have depleted. For the waters we have polluted. For the horrible arms we have in store as if we were Adam's son Cain, jealous of his brother Abel.

Forgive us for the holes we have torn in the ozone layer. We have fought the green environment with bulldozers and spray bottles. And now when we see that we may 'win' the battle, we begin to be frightened of our own success. We pray that nature may resist, because whoever wins out over what has been given us by superhuman evolution, can be the only loser.

We have invented computers, fax machines, and faster ways to save time. But the time we have saved we need in order to earn enough to pay the bills for all these time-saving devices we did not need before. And we have become more stressed and more pressed for time than any generation before us.

We have been despising indigenous people, because they could still marvel at the sight of flowers coming up by themselves from the ground and green leaves growing from the brown branches of trees. They had no microscope to see the chromosomes and genes which we can see. You, our children, must try to see behind what we see, as they did.

We have narrowed our horizons by hiding ourselves behind walls and blinded the heavenly bodies with neon lights. We have worshipped dead things. You must venerate that which is filled with the spirit of life. A robot that could speak when fed with a tape and walk on the floor like a puppet without strings was more marvelous to us than a bird that could sing its own tune and land on a branch in the forest. We learned in church to sympathize with Moses' despair when he came down from the mountain with the Ten Commandments and found his people dancing around a calf of gold. But we treasure our own gold when securely hoarded in banks and prefer the calf roasted as veal.

Look at the flowers as they turn to the sun and the beasts in the stable and the wilderness: they sleep and wake up like you and can move by themselves, select the right food, find their proper mates, and leave for their descendants what they received from their parents, though they have fed on the environment all their lives.

As we did before you, you may wonder forever why you were born and what the purpose of life may be. We thought it was to rebuild the world around us, and we have tried. You may have better success in the future if you try to improve the world on the blind side of human eyes, the world inside man. That is the only part of the world we have left unchanged for thousands of years. What is there is hidden from others and we hide it from ourselves. Inside us is eternity and infinity, and there is where you will find both paradise and hell. That is where God and the devil have their battleground.

Forgive us our ignorance in passing on to you an environment inferior to the one we ourselves received. But do not despair. Take better care of all living seeds. Every one of them has the power to create endless numbers. Birds still sing on this planet and fish swim in the waters. The sun shines on polar bears and elephants. Help to heal the system we have wounded. Try to study and understand the clockwork you are part of.

All that walk and crawl and swim and fly are members of our extended family. And so also are those who can neither see nor hear, stuck in the ground and only able to move so very slowly free of the earth that we think they are motionless and dead, even though we see their leaves and flowers, which were not there some days ago. All that live on this planet – not just the human species – are our relatives. For we have all inherited life from our common ancestors, the single-celled plankton that began the evolution of all life on Earth. All life has a common beginning, whether formed by supernatural creation or by natural evolution. Even evolution, before it began, must have had an input, something creative to set it going in a meaningful direction. If nature created its own evolution, then we must indeed show respect for nature, protect it and venerate it as our superior. If there is a God who created nature through an omnipotent evolution, then respect and venerate his work, which is our environment, and worship your God under any name that fits your faith and your language.

The Seventh Day of God is now. For all human generations.

TRAVELS IN A THIN COUNTRY

A Journey Through Chile

Sara Wheeler

Squeezed in between a vast ocean and the longest mountain range on earth, Chile is 2,600 miles long and never more than 110 miles wide – not a country which lends itself to maps, as Sara Wheeler found out when she travelled alone with two carpetbags from the top to the bottom, from the driest desert in the world to the sepulchral wastes of Antarctica.

This is Sara Wheeler's account of a six-month odyssey which included Christmas Day at 13,000 feet with a llama sandwich, a sex hotel in Santiago and a trip round Cape Horn delivering a coffin. As eloquent, astute and amusing as her first book, *An Island Apart*, *Travels in a Thin Country* confirms her place in the front rank of today's travel writers.

'Notably well-written, perceptive, lively and sympathetic. She catches the elusive character of the Chileans and the unexpected beauty of those remote places which travel books seldom reach . . . very well worth reading'
Daily Telegraph

'Confident and witty . . . not only informative geographically, historically and politically, but also light-hearted and amusing'
The Traveller

Abacus
0 349 0584 7

TIME AMONG THE MAYA

Travels in Belize, Guatemala and Mexico

Ronald Wright

'A superb travel writer'
Observer

Often called the Greeks of the Americas, the Maya created one of the world's most brilliant civilisations. Despite a mysterious collapse in the ninth century and Spanish invasion in the sixteenth, seven million people in Guatemala, Belize and south-eastern Mexico still speak Maya languages and preserve the Maya identity today.

Ronald Wright set out to discover their ancient roots and the extent of their survival. At once a riveting journey and the study of a civilisation, *Time Among the Maya* embraces history, politics, anthropology and literature. Written with wit and wisdom, this is travel writing at its broadest and best.

'Outstanding . . . Wright draws on his experience to make the old Maya as real as the new Guatemalans and it is all delivered with great style'
Sunday Times

'Shows Ronald Wright to be far more than a mere storyteller or descriptive writer. He is an historical philosopher with a profound understanding of other cultures'
Jan Morris, *Independent*

Abacus
0 349 10892 7